Carlitos Luís Sitoie

Geography of marginalised spaces and hidden knowledge

Carlitos Luís Sitoie

Geography of marginalised spaces and hidden knowledge

Perception of the shadows in Haqui (Mozambique) and Macapá (Brazil)

ScienciaScripts

Imprint
Any brand names and product names mentioned in this book are subject to trademark, brand or patent protection and are trademarks or registered trademarks of their respective holders. The use of brand names, product names, common names, trade names, product descriptions etc. even without a particular marking in this work is in no way to be construed to mean that such names may be regarded as unrestricted in respect of trademark and brand protection legislation and could thus be used by anyone.

Cover image: www.ingimage.com

This book is a translation from the original published under ISBN 978-620-2-80428-8.

Publisher:
Sciencia Scripts
is a trademark of
International Book Market Service Ltd., member of OmniScriptum Publishing Group
17 Meldrum Street, Beau Bassin 71504, Mauritius
Printed at: see last page
ISBN: 978-620-2-83165-9

Zugl. / Approved by: The apparent solar movement causes changes in the direction and position of shadows throughout the day and year, influencing the real and the imaginary/cosmology of peoples

*To my parents João Mendonça whom I
eternalize and Amélia Francisco Muchanga.
My nuclear family Stipan, Keccy and
Uchewanda for understanding and
understanding during my long absences.
To my brothers Eugene, Francis, Sonia, Sarah,
Peter, Dinha, Orlando and the two Melites.
Andrielle de A. Marques, my loving and caring
partner.
To the residents of Povoado de Hqui in
Mozambique and the city of Macapá in Brazil.*

ACKNOWLEDGEMENTS

To the Almighty Lord, for life, health and protection.

To my Doctors who I consider them "anthropologists of life", always dedicated, loving and lived. With hands placed in a sign of consideration, friendship and respect, I thank Prof. Dr. Therezinha de Jesus Pinto Fraxe and Prof. Dr. Amélia Regina Batista Nogueira, for their immeasurable involvement in the (re)building of knowledge of the Ba tswa peoples of Mozambique and the Macapa inhabitants of the Brazilian Amazon. My advisors remind me of St. Augustine in his thought: "It is better to love with severity than to deceive with gentleness".

To Prof.ªDra Ivani Ferreira, for the socialization and processing of the acceptance to the PhD Programme in Environmental Sciences and Sustainability in the Amazon.

To Prof. Dr. Henrique dos Santos Pereira, coordinator at the time I joined PPG-CASA, for the help during my stay at the Doctorate, as well as for the acceptance and reception at the Programme. To Prof. Dr. Sandra do Nascimento Noda, for her willing attention, an eternal source of inspiration.

To Raimundo da Silva Nonato (Nonatinho) and his family, for behaving like a true blood brother, friend, father, finally, a human person who treated me as a VIP, welcoming me in everything I needed while attending my doctorate.

To Professors Tatiana Schor, Suzy Pedrosa and Alexandre Riva, for the indication of bibliographical references. To the other Professors and Servers of UFAM that compose the PPG-CASA, by the example of professionalism.

To the residents of Povoado de Aqui and the city of Macapá, who during the fieldwork shared their experiences on the environmental system analyzed from the perception of the shadow of objects and things, providing data necessary for this work.

To the colleagues of the course, Sylvia Forbac, Geize, Gizele, Eloise, Christiane, Eliana Noda, Meire, Wagner, Jhamssem, Raimundo, João Tezza, Aliria de Noronha and Sandro, for teaching me to speak Brazilian, finally for sharing the academic and leisure moments.

To the governments of Brazil and Mozambique for cooperation. To the FAPEAM /MCT-Mz Programme for the scholarship and to the CNPq for the Sandwich Scholarship, which allowed me to improve my research in Lisbon-Portugal.

To Brazilian friends: Jefferson Rolim, Misael Pontoja, Márcia Reis Pena, Edvaldo Meireles; Aneli de Aquino and Antonio Lauro da Silva Marques among others.

My sincere thanks!

SUMMARY

PREFACE

This is an example of the flexibility required of a geographer who, following an approach based on a more engaged and inclusive geography, seeks to find meaning in the daily life of local communities where knowledge is shared and passed on from generation to generation without being systematised into specialised treatises as in Western science. Having worked in African environments, Milton Santos was one of the promoters of the need to value local creativities, lower circuits of the economy, fruits of knowledge and non-standard, improvised solutions. Boaventura de Sousa Santos also encourages the inclusive approach with his conceptions of the epistemologies of the South and the ecology of knowledge, justifying what in Mozambique we have been calling inclusive geography not only of people with special educational needs but of marginalised spaces and hidden knowledge whose manifestation differs from the standardised form of Western knowledge which has always claimed the status of universality and exclusivity. However, this inclusion always implies a symbiosis of cross-fertilisation with the formal knowledge accumulated by humanity throughout its history, as did the author of this work who, in dealing with the shadow, sought his role in the history of geography from classical antiquity.

The shadow being often connoted with darkness, darkness, sin and curse, paradoxically constitutes, according to this book, the element that very early in history made it possible to understand light. It exists because of light and its interception with small objects as big as an ant, such as the stars Earth, Sun and Moon, when they interpose themselves causing eclipses.

A comparative study between communities in Mozambique and Brazil shows the strength of historical-geological destiny and political history, of the dialectics between separation and union from which diversity develops not as a weakness but as a wealth of complementarities. Africa and South America are both southern continents from the same supercontinent, Gondwana. Colonisation has brought us a common language that allows easy communication and learning of different but also complementary knowledge and practices. From the colonization the cashew tree of Brazil passed to Mozambique, an emblematic plant growing along the North to South Coast having deep roots in the culture and whose phenology constitutes the calendar of many communities such as the village of Aqui, which was the object of study of this work, the cashew tree and other trees that project shade, function as a calendar and clock, influencing the rhythm of daily activities.

The author shows us the versatility of shadow and the discovery of the ways to use it. Several sciences study how to use it for the economy of light to ensure a certain microclimate or for the maximum use of light. It is in fact an interdisciplinary area of agronomists, astronomical ecologists, geographers, architects and landscapers that interests all mankind, paraphrasing John Paul II. Whole Man considering body and soul. In the Chi-Changana language of Southern Mozambique shade is also synonymous with home, a livable, welcoming place. The selection of shade trees for the living space planted or found is an effort to make the house more welcoming.

At a time when efforts are being made to achieve the Objectives of Sustainable Development, this work deals with an aspect that contributes to the defense of trees and the construction of a new look on trees in our geographical environmental education that has become more inclusive of its potential that has been simplistically only connoted with the production of oxygen. Who knows if in the monument to be erected in the line of the capricorn tropic in Mozambique it will not be necessary to place a tree that also imitates a palhota with a conical roof, a school a court, a common house, a home where the problems of the people are solved under a good shade.

Having visited the State of Amazonas, precisely at the Federal University of Amazonas (UFAM), at the University's invitation, for the Health Geography Congress, I noticed that the university campus was equipped with leafy trees and good shade, I take advantage of this manuscript to thank you for the invitation and pleasant welcome. I congratulate the author Carlitos Luís Sitoie for this very promising work, hopeful that it will bring a new impetus to the teaching and research of Geography and Environment in Mozambique and Brazil in an inclusive approach of diverse, hidden, inclusive, but fruitful knowledge.

Zechariah Alexandre Ombe
Associate Professor of Geography, Coordinator of the Doctoral School
from UP Maputo, Mozambique

PRESENTATION

Faced with the challenging context experienced by all peoples who still use everyday knowledge and practices for their experiences, new approaches coming mainly from non-Eurocentrist authors have sought to bring their countries' socio-spatial, cultural, political and economic thematic approach closer to the global context. Trying to break concepts of tutelage and assistance that historically marked the scientific approaches in favour of the European current, harming the knowledge and practices experienced by all those who did not fit the rule of the scientific method.

The book in question constitutes one of the forms of inclusion of marginalised spaces and knowledge, that is, those knowledges and practices that have always been relegated to the background by so-called scientific knowledge. Shadows have never enjoyed a good reputation, understood as the dark, obscure, frightening places, representatives of bad things, are approached in this book as phenomena and categories of analysis that represent good things, those of good living, synonymous with comfort and well-being.

As a result of his doctoral thesis, he seeks to understand how the line of the capricorn tropic in Mozambique and the equator in the city of Macapá in Brazil influence the daily lives of the residents of the places studied. For the book's realization, a methodological approach was developed to guide the comprehensive conception of a phenomenological geographical and ecological perspective, which allowed integrated readings of two distinct but similar places (Macapáurbano and Haqui-rural).

The objectives of the book focused on understanding how the perception of the residents of Haqui and the city of Macapá is articulated from the shadow of solstice and equinoxes.

The visual material aimed to illustrate texts with graphic language through tables, diagrams, maps, photos and drawings.

Making this publication possible is one of the ways to contribute to the interaction between marginalized spaces and knowledge, with society in general, by incorporating these places, knowledge and practices, into the global context, integrating them as material resources of the patrimony of humanity.

Carlitos Luis Sitoie

INTRODUCTION

The book seeks to uncover the symbolic relationships impregnated in the residents of two geographical spaces: Haqui village in Massinga district, Inhambane province in Mozambique and the city of Macapá in Amapá state in Brazil. Based on geographical and environmental science in its approach based on Tim Ingold's ecological paradigm. The publication of the book appeared among several motivations, from the need to visualize spaces and knowledge "hidden" or considered folklore, for manifesting itself different from the standardized form of western knowledge, which uses the rule of scientific method through writing.

Based on oral narrative, several peoples, such as the *Batwas* of the village of Kamuhaqui, who do not use writing, claiming exclusion from their geographical space and knowledge, who are ignored by science on the grounds that they do not have the status of universality and exclusivity.

Despite this mistake, these peoples, driven by the desire to organize units of time, know and orient themselves in geographical space, using methods and techniques that provide thermal comfort, demanding human capacity for environmental perception, usually made by means of scientific technologies and disseminated in the media in recent decades. These population groups use their analogical capacity[1] of perception to interpret variables of the environmental system, such as solstices and equinoxes visualized through shadows, constituting one of the most accessible techniques for the population of all social strata.

The shadows projected by (objects and things)[2], allowed knowledge about the local geographic space "[...] when necessary a defined course. Moreover, when fishing, hunting and trade involve great distances, the need to know the way back and forth is obvious [...]" (MILONE, 2003, p.12).

The choice of shadows as the object of study was not random, as Vansina (2010) said, the researcher should begin first with the ways of thinking of his people before interpreting the practices. Belonging to the *Ba tswa* ethnic group, I learned from the elders that shadows play an important role in the organisation of social, economic and political structures. Allowing the organization of annual cycles of activities, guiding choices of places to bury the dead, identifying animal shelters to hunt, looking for fruit and roots, running away from dangerous animals, adapting to the alternation of the dark light and the changing seasons.

---

[1] Innate capacity of human beings through their organism to interpret/analyze environmental variations through their sense organs: ears, smell, taste, sensation of heat, cold, pain or pleasure without resorting to devices or technologies.

[2] Concept of Rhodes (1996), where objects correspond to natural environmental variables (e.g. flora and fauna) and things to elements resulting from human construction (e.g. furniture).
According to Crátilo (387c, 388a), they are instruments, names or symbols used to inform about things. In this case, all the variables of the environmental system that project shadows, are part of the things that interest our study.

The option for the shadows of the tropic in the book, and in particular of the village of Aqui (in Mozambique), started from the observation of the disputes involving the district government of Morrumbene and Massinga, which each forced the installation of signs signalling the passage of the imaginary tropic line of Capricorn over their district. Implanting milestones in a random place, that is, a place that did not correspond to the conventionally recognized geographical coordinate. The dispute began in 2008, when the then President of the Republic of Mozambique, Armando Emílio Guebuza, in his open governance speech, explained the tourism potentialities that the Tropic of Capricorn line and its phenomena could add to the region. This speech impressed the leaders of the two neighbouring districts (Morrumbene and Massinga) to try to signal the crossing of this line in their territory. First they took the landmarks that had been fixed in Massinga since the 1980s (80s) and passed them to the neighboring district (Morrumbene); in 2014 in January, the signs returned to Massinga after two movements. This mobility of the landmarks, both south and north, along the National Road (EN1), which connects the south and north of the country, aroused curiosity, questioning, and a desire to understand the impact that this imaginary line has on the lives of the residents who live around it.

The geographical scope (MacapáAqui) results from the need to establish a comparative study between communities influenced by the zero shadow phenomenon, seeking to study the daily life of these two places where the research took place, very distinct but with similarities.

Distinct study areas due to their geographical location: the city of Macapá crossed in the middle by the imaginary Line of the Equator in the Brazilian Amazon and the rural village of Aqui situated in the Province of Inhambane in Mozambique, crossed by the imaginary Line of the Tropic of Capricorn. Similar because they belong to the tropical region where the phenomenon of zero shadow occurs[3,] during the equinoxes and solstices respectively.

From Cárdenas *et al* (2003), Rice and Greenberg (2004) it is possible to deduce that shadows, in particular of trees, have the power to increase the ability to attract greater diversity and abundance of wild birds and mammals, which can be more attractive to beneficial insects such as honey bees. Making it important to learn the trajectory of these shadows in order to plan hunting seasons, honey harvesting, as well as the appropriate periods to take better advantage of these shadows in various activities in rural and urban areas. In the Povoado de Aqui, the appropriate times for weeding, sowing, weeding, harvesting and the organisation of festive dates are planned according to the mobility of shadows.

The festivals often coincide with the position of the shadows "[...] solstices or equinoxes [...]" (ALVES, 2006) time of festivities in the village of Aqui and the city of Macapá. In Aqui, the

---

[3] According to Alves (2006), when the sun is halfway to the pin, it is when it has already made in its apparent movement a trajectory corresponding to the solar midday and the shadows overlap on the objects/things during the solstices and equinoxes.

festivities symbolise food abundance, community joy, peace, tolerance and truce among the inhabitants.

For the *Ba tswa* ethnic groups located in Inhambane province, the planting and choice of tree species to be planted are selected based on the type of shade they will offer throughout the year. Serving as a shelter from the sun, a symbol of family division according to sex, age, judicial and sacred power. Because in the domestic yard there are several shadows representing each one the place of men, women and children. Others symbolising a legal institution, because it is the place where the "*Madodas*"[4]meet, or the shadow of the "*nhamussoro/nhanga*"[5].

The shadows are loaded with thoughts and meanings to the point that some public institutions in Mozambique present a shadow tree, used for consecration of rituals of power or evocation of spirits of the deceased, venerated as important figures to control peace and war within the living. Examples of officially constituted shadows are: that of the Universidade Pedagógica de Moçambique - Delegação de Massinga and the Sombra do Comandante, at the Escola Prática da Polícia in Matalane, used to invoke the spirits of the deceased which guarantees security and integrity within the institution. Or they serve as a place to visualise the power of hierarchical superiors in relation to subordinates (in the case of the Matalane Police Practice School, which only officers have the right to attend that shadow). The choice of place for courtship or sexual relations also follows the type of shadow. Surveys made with elders of the village of Aqui refer to the change in the trajectory, position and size of the shadow objects, affecting the organization of their socio-environmental structures.

The sunshade is part of the "[...] position astrology [...]" (RODRIGUES *et al*, 2010, p.24)" used for geographical orientation indicating the cardinal points and serves for time measurement. Taken by Harvey (2001) as the best way to explain environmental issues, especially when analysed through the timeline.

Starting fromfonso (2006), Rodrigues Júnior (2012), the Earth's translation movement influences the timeline, which can be recorded by controlling changes in the position and shadow size of objects, things and people throughout the year. Facilitating the identification of environmental particularities related to natural events or important moments of everyday life, especially in traditional communities, lacking modern technologies, without access to scientific knowledge and the media, but which use their knowledge to live and produce knowledge.

Knowing that they are often made invisible by scientific knowledge because they are considered folklore, there is a need to analyze the actions of these people, trying to understand how they adjust the calendar and other actions related to the environmental system. This is

---

[4] People of the adult age group, considered in the village of Aqui, as "knowledgeable" of the techniques of reading and interpreting the shadows, including sufficient maturity to contribute to the solution of various socio-environmental problems in the village.

[5] Healer or person with the ability to invoke the spirits of the dead/forefathers in *Xitswa or Tswa* language.

relevant because it allows scientists to know how the public reacts to climate impacts or initiatives, because these reactions can mitigate or amplify impacts. Bordet *al,* 1998.

Supported by Afonso (2006), Alves (2006) & Scandiuzzi(2000), it can be said that today there are still people mainly living in the rural area, such as the ethnic groups of the African6 tropics and Brazilian indigenous people, who use the shadows for various purposes, such as, for example, to guide their villages even today with new technologies such as GPS. That is, people who find in shadows projected by objects, positive aspects mentioned by Nowaket *al* (2001), D'amral (2003), Frota (2004), Dias-Filho (2006) and Belchior (2014) as being the decrease of solar ultraviolet radiation, contributing to the decrease of health problems associated with the increase of exposure to such radiation, such as cataracts and skin cancer. Allowing these people to establish natural cycles, social, economic and political organization.

The same authors list the following negative effects: shadows when used for tethering animals are places of total loss of soil cover and susceptible to compaction and erosion due to trampling and soil exposure, constituting places of greater accumulation of faeces and urine, reducing soil fertility due to waterproofing.

It should be noted that as the angle of inclination of the earth's elliptical in relation to the celestial axis changes, it causes changes in solar irradiation and in the behaviour of shadows of objects and things on the earth's surface, affecting the environmental system of people who use the shadow for time counting and geographical orientation.

For the collection of data that gave rise to the book, it was assumed that the variation in the angle of the solar elliptic influences the climatic variations that impact the daily lives of the inhabitants of the places crossed by the Equator (L.EQ) and the Tropic of Capricorn (T.CAP). That through the zero shadow of the solstice and equinoxes, they find strategy for processes of environmental adaptability. Reorganizing social, economic, and political structures. Seeking interpretations of environmental variables that correspond to the interrelationships of knowledges considered folklore and scientific knowledge, from the interpretation of objects, things, subjects and/or phenomena that represent realities that are in the process of transformation.

Because it was produced in this way, the book took into account and articulated the lived knowledge, accumulated by the experience of those who experience this phenomenon, as scientific knowledge built from data obtained based on the information of those researched.

---

[6]Trajectory of the earth in relation to the apparent solar movement, from the tropic of cancer to the capricorn and vice versa. "[...] this area comprises 47° South/North latitude from the equator, the equivalent of [5],405 km, based on the equivalence of 111 km for each degree of latitude. In terms of longitude, its area covers 360° along and around the equatorial line, which represents an area above 40,000 linear Km at zero latitude (0°) [...]' (BENCHIMOL *et al*, 2002, p. 137). Determining the "[...] solar year as the period corresponds to two consecutive passages of the sun's apparent movement through the equinox at the equator. For this reason, the elaboration of annual calendars in many societies has obeyed the tropic year, making it possible to follow the cycles of each species from the heat regulated by the sun throughout the year [...]" (RODRIGUES JÚNIOR, 2012, p. 40).

Making surveys of shadow perceptions in two places located in different continents, countries, that is, distant geographic spaces, which can be used by students and teachers to generate knowledge, minimizing the deficiency about the programmatic contents of Mozambican and Brazilian teaching, referring to the didactic unit that deals with geographic coordinates and tropicology, climatology and thermal comfort.

It will be possible to understand, with this work: how the perception of the inhabitants of Aqui and the city of Macapá is articulated from the shadow of solstice and equinoxes; identification of objects/things used to control shadows projected throughout the year in Aqui and the city of Macapá; how are [7]the social representations of groups that have been studied from the mobility of the shadow of objects, things and social actors; carta/diagram solar of the village of Aqui and the city of Macapá; the knowledge and practices related to the production of life, based on the mobility of zero shadow in Macapá and Aqui.

In order to establish the relationship between the two places enunciated in the work, the idea was born that the zero shadow represents the equinoxes and solstices, being these environmental variables that dynamize organizational structures of human social groups. The solstices and equinoxes are elements that facilitate social, economic, political organization and temporal division through the seasons, constituting important elements for establishments of vital cycles and thermal comfort.

---

[7] The representation that a group elaborates on what to do to create a network of relationships between its components by defining the same specific objectives and procedures. The first step in social representation is the elaboration under induction of a task that does not take into account the organization of social behaviour, but rather the functional organization of the cognitive reality of the group. These are the ways of thinking and interpreting everyday reality and fixing their positions on events and objects. A social that manifests itself through the context in which the groups live, practical knowledge that gives meaning to everyday life. A process that establishes a relationship between the world and things (SÊGA, 2000, p. 128).

CHAPTER I

13

SHADOWS IN EVERYDAY LIFE

The chapter seeks to address the shadows projected by (objects and things)[8] as phenomena that allow the understanding of geographical space. In addition to allowing geographical orientation, when fishing, hunting and trade involve great distances, making it easier to know the way back and forth.

The reading and interpretation of shadows to organize everyday structures has always been one of the greatest concerns of human societies. This is because they provide thermal comfort, improve the quality of meat in animal production and because they are used in the production of low photoperiod cultivars. In architecture, medicine, philosophy, geography, mathematics, physics, drawing, art, among other areas of school knowledge, they occupy a prominent place being subjects of teaching-learning contents. Knowledge about the spherical shape of the Earth had its presuppositions in the perception of shadows, and even today many people use these phenomena to guide shading facades in the construction of their homes. They are also used for the tracing of paths and azimuth on hiking trails, in the determination of distances from their numbering along the way, they are also used for the production of calendars or annual cycles, among others.

The residents of Macapá city, Amapá state in Brazil and Povoado de Aqui in Massinga district, Inhambane province in Mozambique, perceive and use their life experiences to take advantage of shadows. There are similarities and differences that reflect the social and environmental particularities of the two places elucidated. The remarkable moment in the trajectory of shadows corresponds to the moment when objects and things overlap with them, in the phase called equinox, happening in the city of Macapá twice a year (March and September) and in the December solstice in the village of Aqui, occurring once a year.

All these phenomena were researched as fieldwork, in a theoretical approach combined with the ecological paradigm of TimIngold Thought, with observations and interviews of social subjects.

In Macapá, the interviews were made during the equinoxes involving the residents participating in the ritual of the sun passing through the obelisk of the Marco Zero monument. In the village of Aqui, meetings with the *madoda* or *elders* were prioritised. The socio-environmental organisation in the city of Macapá obeys the water and drought equinoxes. While the perception of the shadows of the inhabitants of Aqui is impregnated with concepts related to flowering, foliar abscission, position astronomy, sundial, photo periodism and the establishment

---

[8] Concept of Rhodes (1996), where objects correspond to natural environmental variables (e.g. flora and fauna) and things to elements resulting from human construction (e.g. furniture).
According to Crátilo (387c, 388a), they are instruments, names or symbols used to inform about things. In this case, all the variables of the environmental system that project shadows, are part of the things that interest our study.

of seasons, allowing the elaboration of calendars or natural cycles based on the mobility of shadows for the practice of various activities.

1.1 City of Macapá - Brazil

Macapá, capital of the State of Amapá, was incorporated into the studies that elucidated this work because it is located in the geographical area corresponding to the tropical area covered by the apparent solar movement during the translation of the earth, which moves from the Tropic of Cancer to Capricorn and vice versa, crossing the city twice, impacting on the daily lives of the residents in the form of equinoxes or zero shadow, celebrated at the Marco Zero monument twice a year. The crossing is equivalent to:

> [...] solar year, as the period of two consecutive passages of the sun's apparent movement through the equinox in Ecuador. This solar mobility has always been used to plan the elaboration of several annual calendars obeying in many societies the tropic year, making it possible to follow the vital cycles of each species from the heat regulated by the sun throughout the year [...]. (RODRIGUES JÚNIOR, 2012, p.40).

The city is in the Amazon, which covers nine Latin American countries, being the Brazilian one comprising the states of Acre, Amapá, Amazonas, Mato Grosso, Rondônia, Roraima, Tocantins, Pará and Maranhão, in its portion west of the Meridian 44° (MAPA 1). Constituting an environmental system with its own organization and interactions, it reflects that of unique landscapes, cut by large rivers and inhabited by diverse human populations. These populations are responsible for building, from the available environmental resource, places with societal structures or particular points of living, which behave with an organization and their own interactions, revealed from attitudes that represent their social groups (BRAZIL, 2007). Following the example of these societies, one can cite the inhabitants of Macapá, a city in the middle of the world due to its strategic location in the geographical coordinates ("00°02'18 84") under the imaginary line of Ecuador.

Map 1 - Macapá's geographic framework in Amazonia

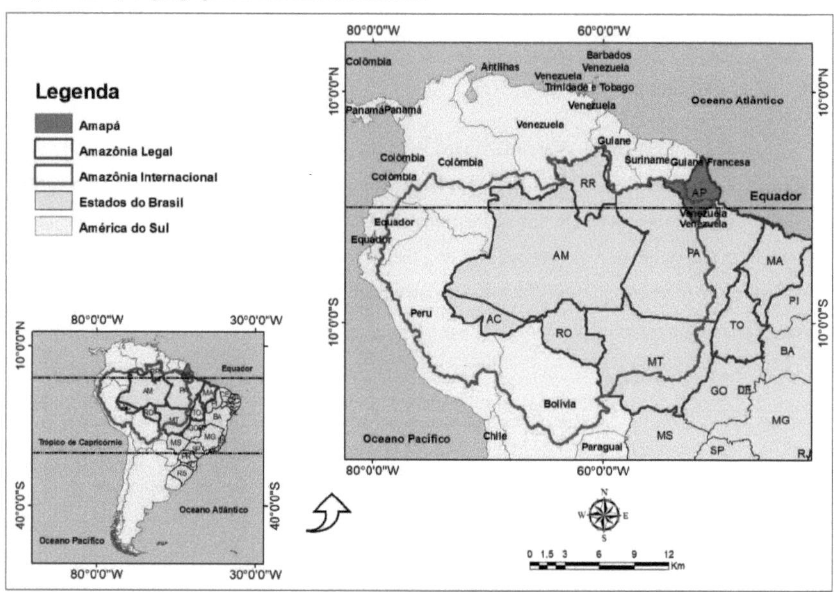

Source: The author (2018).
Note:Using HQIS 2.18.15 *las palmas*.

Macapá belongs to the state of Amapá, situated on the shield of the Guianas,[9] in the northern region of Brazil, in an area of 6,407.123 km², with a population of 465.495,000 inhabitants estimated by the IBGE (2016), including the city in the Amazon Biome, which is made up of a great diversity of ecosystems and forms of relief that reach fourteen (14) metres in altitude in relation to the average sea level, making up "[...] the only state capital cut by the Equator and bathed by the Amazon River, but has no road or rail interconnection with other capitals of Brazil; only air and water [...]" (TORRINHA, 2015, p.10). Drawing an urban perimeter corresponding to the following geographical coordinates 00°8'37"S; 51°16'18 33"W; 00° 10' 9" N and 50°56'45"W, on the banks of the Amazon River (MAP 5).

---

[9] The Guiana Shield corresponds to the central and western sections of Amapá. In the southern and northern portions, the cenozoic sedimentary deposits (coastal plains) extend, about 25% of the state, formed by fluvial and fluvial marine deposits, and where Macapá is located.

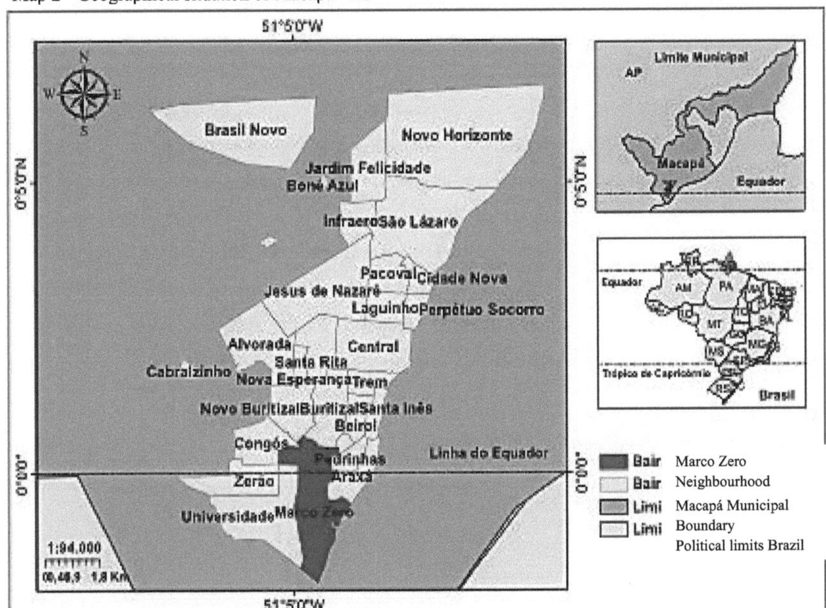

Source: The author (2018).
Note: HQIS Resource 2.18.15 'las Palmas'.

Macapá is also known for the nickname Middle of the World or Marco Zero city, given the geographical situation over latitude zero (00°) degrees that divides the municipality into two hemispheres N/S, gaining metaphorically this nomenclature that attributes a symbolic and significant character to the city including the state of Amapá.

The location over the flowing Amazon River, in a region crossed by the imaginary line of the equator, means that the city and the state strategically receive a greater incidence of solar rays and water vapor, making the equatorial climate hot and humid. It is characterized by precipitation that is subject to seasonal variations due to migration from the Intertropical Convergence Zone (ITCZ), charged by convective air masses, associated with the confluence of trade winds in the region of low atmospheric pressures, causing the formation of convective clouds over the equatorial Atlantic ocean, which may spread towards the Amazon by the eastward runoff, according to Figueroa and Nobre (1990), Paiva and Clarke (2011), Marengo and Nobre (2009) and Tavares (2014).

The "[...] region often suffers from anomalies in climatological averages due to the influence of extreme climatic events, resulting from large-scale variabilities, such as the El Niño and La Niña phenomena [...]" (NEVES, 2014, p. 2). The city's climate is equatorial with a short dry season, in the months of October and November, receiving the AM (Koppen) and Geiger classification, Equatorial Humid (Strahler), B3 (Thorntwaite) and Equatorial 1b (IBGE, 2016).

The rainfall averages around 2487mm per year, reaching extreme values during the March equinox, which is why it is also known as the water equinox, because of the increase in rainfall levels in the city, this increase starts in January with a peak in March (407.2 mm/month).

In September there is the equinox of droughts, so called due to greater sunshine and lower water levels, reaching the lowest rainfall of the year in September and October (35.5 mm/month). There is a difference of 370mm between the rainfall of the driest month and the rainiest month. Precipitation includes about "[...] 169 days of heavy rainfall during the rainy season (December to July) and 196 days of low rainfall, which is considered the dry season in the city (August to November) [...]" (IBGE, 2016).), as can be seen in Graph 1.

Graph 1 - Macapá rainfall variation

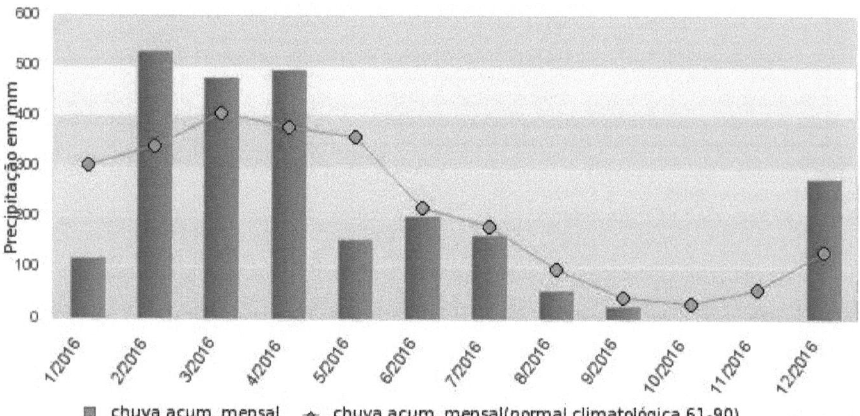

Source: INMET (2016).

In addition to presenting the variations in rainfall in the year two thousand and sixteen (2016) from the monthly accumulated rainfall representing the climate characteristics of the city of Macapá, the graph represents a graphic language that allows us to analyse how everyday practices, which depend on atmospheric precipitation for family agriculture and navigation, for example, can be structured or organised, to the detriment of the pluviometric characteristics of the place. That is to say, the graph allows one to know the months of high rainfall (February, March, and April), months when it rains normally (June, July, and December), and periods of low water surplus (August, September, October, and November), with these rains normally occurring in the early morning, morning, and afternoon.

Analysing rainfall variability, it is remarkable that phenomena related to geographical location, demarcate moments of rainfall variability, with the highest values being reached at the flood/water equinox, while the lowest rainfall rates coincide with the festival of spring or droughts in September. That is, the low or high rainfall rates are related to the movement of the thermal equator during the seasons, the Intertropical Convergence Zone. The seasons happen because of the translation movement and inclination of the Earth. From this perspective, these phenomena are influenced by the crossing of the equator through the city, playing an important role in marking the beginning or end of the seasons. This means that the equinoxes symbolize rainy and dry seasons, facilitating day-to-day planning, as residents use their experiences, for example, to manage flooded areas, such as the hangover,[10]Indian lagoon, Beirol, Muca, Burtizal, University, Novo Horizonte, among others, which need in the rainy season, better organize their stilts, floating or abandoning their residences located in the lower area, moving to the higher ones.

Studies such as Varejão (2001), Mendonça, and Danni-Oliveira (2007), argue that hangovers, despite being flooded areas, play an important role in the local microclimate, providing moisture to the atmosphere, and serving as a sink for rainwater, as they are in depressions connected to the Amazon River.

Fishing and navigation, mainly over the mouth of the Amazon River, reach high peaks during the drought equinox, the same period in which thunderstorms occur massively, while fishing and navigation activity registers low values, during the water equinox phase. It is at the water equinox that pororoches occur[11], accompanied by heavy rains during high tide, particularly at high tide or new moon, causing flooding.

Due to its geographical location, Macapá receives more days of sunshine than the cloudy ones, making the city hot and uncomfortable almost all year round. This is due to the geographical situation of the city on the imaginary line of the equator, which receives a higher incidence of solar radiation compared to other regions of the earth, located at latitudes greater than zero (00°) degree, see Graph 2.

In relation to other regions distant from the equator, because the whole region in the imaginary line of Ecuador receives more incidence of solar rays, that is, it is not exclusive the case of Macapá. Thermal discomfort, too, must be combined with urban densification, few wooded areas, soil sealing, street layouts, etc.

---

[10] Regional name given to flood plains or basins, influenced by tidal, river and rainfall patterns.
[11] Large waves of water, which take place at the equinox of the waters under the mouth of the Amazon River in Macapá and Belém, emitting huge bursts.

Graph 2 - Variability of sunlight and temperature in Macapá

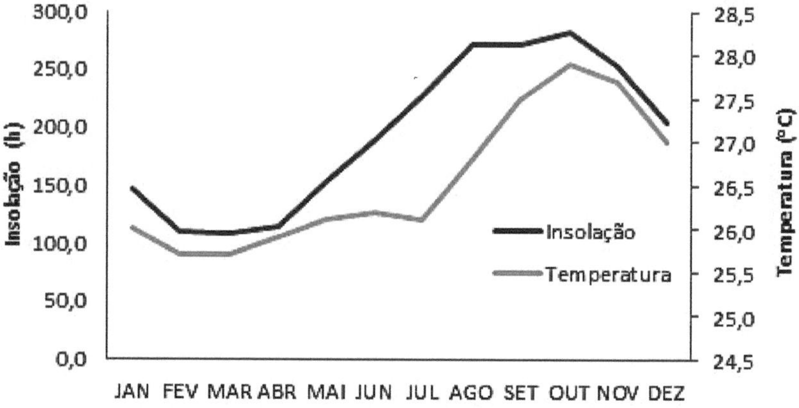

Source: INMET (2000).

In this perspective Varejão (2001) corroborates by saying that the rise of the sun at noon, measured in Macapá, varies little around 90°, since the city is crossed by the Line of Ecuador. Thus, the amount of energy that reaches the surface, per day, varies between 34 and 36 MJ/m2, depending on the time of year.

Thus, the

> [...] higher values occur during the equinox months, when the sun passes vertically from the Equator in March and September. This great amount of energy that reaches the surface helps to keep the temperatures always high in Macapá. And due to the high humidity throughout the year, the thermal amplitude is very small, not exceeding 10°c. The maximum temperatures are between 31°c and 33°c, but the maximum temperature during a day can reach 40°c. Between August and October the highest temperatures of the year occur. The lowest average temperatures occur in March between 25 and 26°. The average maximum temperature of the warmest month occurs in October, reaching 32.6°c while the average minimum temperature of the coldest month occurs in July, with 22.6°c. This minimum temperature, which occurs around 06:00HL (local time), is related to the absence of cloudiness and low relative humidity at this time of the year, allowing the infrared radiation emitted by the surface not to be absorbed and reissued by the clouds or humidity (greenhouse effect) [...]. (NEVES, 2014, p.7).

From the interviews with the social subjects it is possible to affirm that the thermal variability of the city is perceived, based on the experiences lived, that through them, the Macapatians change their work routines, looking for adjustments to the detriment of the heat that is registered almost all year round. The humidity indexes are high in (February - March) and low in (September - October), registering, variations of the annual averages and monthly maximums, in the instant that the equinoxes happen. That means that the lowest annual temperatures are

20

registered in the period between February and March, those minimums observe temperatures that are not cold at all, being very hot, reaching about twenty-five, point six (25.6°c) degrees centigrade.

Another climatic element that acting on the geographical situation dynamises the daily structures of Macapá is the circulation of air masses taking into account direction and speed. Knowing the predominance of wind for a certain locality is of great importance, since it influences agriculture, transport, and especially air and sea navigation, construction, dispersion of pollutants among others (LYRA, 1998), so the wind data are important to elaborate wind forecast and production of hodographs[12].

The intensity and direction of air movement are of simple measurement, which can be done using everyday practices such as controlling the movement of tree leaves and other objects, including through the sensation from sense organs when the air comes into contact with the human body. The experience of the senses of the population can be verified in the statement when the air "[...] touches my body, I feel cold with river water, so I know it is breezes from the Amazon River, but when it is hot, I know it is coming from the other place far from the river [...]" (verbal information)[13].

The direction and intensity of ventilation are important indicators for analysing the distribution and frequency of winds throughout the city of Macapá, in order to predict moments associated with weak, moderate, strong winds and storms (CHART 3).

It is also possible to analyse thermal comfort, especially when the geometry of the city, position of houses and streets are known. In other words, the design of ecological meshes, or the allocation of objects, must be done in a way that allows the city to have good air circulation, airy and environmentally comfortable.

Neves *et al* (2014) allow us to say that the greatest variations in the direction and speed of the winds throughout the city are recorded in the last quarter of the year, specifically in October, November and December (OND), with 10% in the direction N (north), 28% in NE, 49% (E) and 7% in the direction SE (southeast). Conjugating this observation with the interviews of the social subjects, it can be said that, in spite of the weak and moderate winds registered in February, March and April (FMA) by Macapá, this time of the year is not the best for navigation, going to the beach, or carrying out other leisure activities on the river, because it is the time of the equinox of the waters in which the pororocas occur.

The months of (OND), correspond to the equinox of droughts, a less rainy period, with strong winds, and gusts may occur more frequently, due to the differential heating in the surface layers of the earth, which is high, causing high temperature gradient, consequently increasing the

---

[12]Lines or graphic representations that represent trajectories, speed or direction of the winds, and can be plotted in the form of isotopes and isogons.
[13] Subject L.S.M., 31 years old, city of Macapá, Amapá, Brazil.

turbulence of the air masses, i.e., the high thermal amplitudes increase the difference in atmospheric pressure between the continent and the river that bathes the city, the Amazon River, intensifying the river breeze.

Chart 3 - Wind Hodographer[14] in Macapá (2008-2014)

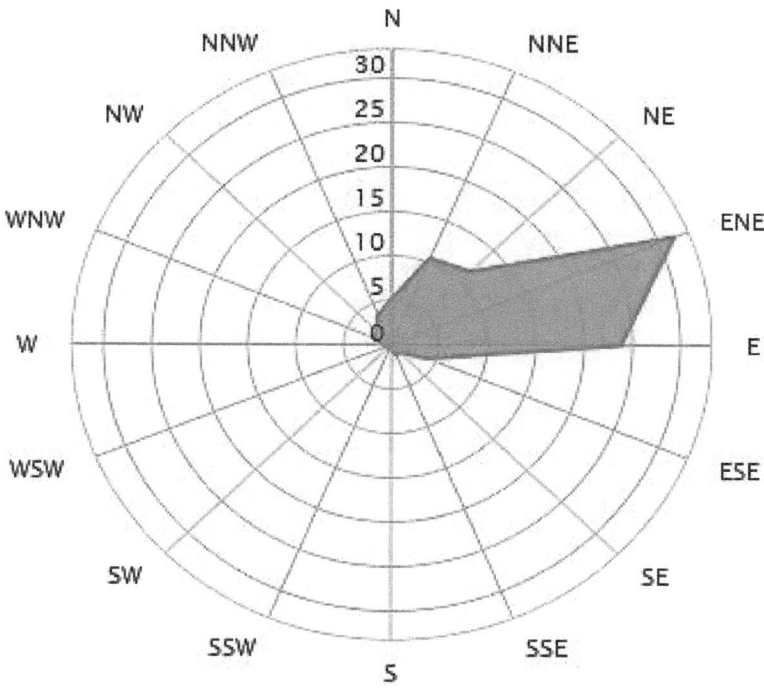

Source: Windfinder (2013?).

The winds in the city of Macapá are from the Northeast (NE), with variations between the East (ENE) and the East (E). The intensity varies throughout the year, but in general the city is ventilated with weak to moderate winds (0 to 25 m/s), with greater predominance in the direction E (East), bombing breeze-type ventilation, loaded with humidity, transported from the edge of the Amazon River that covers the city centre.

According to Figueroa and Nobre (1990), the months with higher relative humidity are of less thermal comfort, due to the saturation of the air from the water vapour that inhibits the evapotranspiration, besides giving the sensation of muffled time and thermal discomfort to the body.

---

[14] Tool or graph that continuously records the intensity and direction of the winds.

From the point of view of the use and occupation of geographical space, Tavares and Tostes (2013) state that Macapá is a new city resulting from a troubled, disordered and emergency-built process of evolution according to the demands that arose. Throughout this growth, some urban planning strategies have emerged, through master plans that were conceived under the influence of important events related to the region, in an attempt to respond to demographic demand without respecting environmental details, such as the construction of ecological nets of thermal comfort of the city. Designing roads, squares and lots oriented in a north-south and east-west direction, following the line of the Amazon River bank.

With the implantation of the Industry and Commerce of Minerals (ICOMI) in 1947, in Serra do Navio, it begins the expansion following the orthogonal mesh using areas called capoeiras occupying areas of flood plains, constituting the neighborhood of Santana, Lagoa dos Índios, beginning the occupation of some areas of hangover, still incipient, (Portilho 2006).

According to Brazil (1969) the occupation of the current southern area of the city began in 1969, based on the preliminary economic diagnosis of the urban areas of Amapá, prepared by the Ministry of the Interior together with the Department of Economic Studies (DESEC) and later with the intervention of the João Pinheiro Foundation, in 1973, which deduced the need for expansion to:

> [...] southern zone of the city along the JK highway (Macapá/Fazendinha road), as well as the expansion zones to the northern sector, where the districts of São Lázaro, Renascer, Jardim and Infraero I, Pacoval district, are currently located, and in the southern direction, towards the university nucleus of the Federal University of Pará (UFPA). (PEREIRA, 2003, p. ?).

Néri (2004) and Portilho (2006) disagree that the master plans were useless because the occupations were carried out in accordance with the invasions motivated by the population explosion towards the 3° BIS (Ministry of Defense-Army), with the appearance of the Alvorada neighborhood, São Lázaro on the banks of BR 156, Perpétuo Socorro, Baixada do japonês and Jardim Felicidade.

In the 1990s some political factors contributed to the migration towards the state of Amapá, having as main attraction factors: the transformation of the Federal Territory into the state of Amapá (1988) and the creation of the Free Trade Area of Macapá and Santana (1991).

Pereira (2013) states that the transformation of the Federal Territory into the State of Amapá and the creation of the Free Trade Area of Macapá and Santana, encouraged by the financing lines of the Caixa Econômica Federal (CEF), have driven a strong migratory process and a high rate of occupation in the centre and sub-center districts. Intensifying verticalized construction (buildings), in neighborhoods allotted as: Buritizal, Laguinho and Santa Rita, the group Laurindo Banha, Cabralzinho and Cap Azul Jesus de Nazaré, Pacoval and São Lázaro (1984); Jardim Felicidade (1985); Novo Horizonte (1994); Cap Azul and Brasil Novo (1997);

Novo Horizonte II (1994); Infraero (1997); Liberdade (1999), Canterbury (2011), appearing tall buildings such as the Cathedral Residence, the Vex complex, the Twin Towers of the Icon, the tower diFirenzze in the middle north, among others.

For Monteiro; Oliveira (2013) the construction of buildings has worsened the climatic conditions of the space, serving as a barrier and reducing the average speed of the winds, since they work as a windbreaker in the natural ventilation process, corroborating with the alteration of the ventilation conditions, boosting the heat gain by changing the roughness, the shape of the relief and the sealing of the soil.

For Nunes (2012) the verticalization generates shadow projection on the light and contiguous houses of the neighborhood, with partial loss of natural sunlight. This model increases the humidity in the internal parts of these houses, proliferation of fungus and grass in the woods and cabinets, asthmatic and broncho-pulmonary diseases, among others.

Moreover, the tall buildings around the edge of the Amazon River associate the accumulation of humidity and mildew in the internal part of the houses with solar obstruction, especially during the equinox period, since it is at this moment that the angle of solar slope is straight, projecting shadows superimposed on the buildings themselves, remaining dark because it does not receive sunlight inside the houses.

Associating the difficulty of the penetration of solar radiation inside residences causing stress to the city residents, who complain about the lack of radiation or sunlight in the rooms, associated with the infiltration of water inside, making the interior of the residences muffled, and the damp smell of clothes that become fragile tearing fast. This situation is influenced by the construction of high-rise condominiums that do not allow sun penetration into the interior projecting shadows that make the place always dark and without sunlight, as one resident claims:

> [...] these buildings only came to spoil the view to the edge of the city, which was very visible from any point, besides, in these buildings children no longer play in the sun, play ball and other games for body resistance, it is only in the air conditioning, fan, sweat that it is never good, so children of today are weak and obese. We used to catch the sun playing even at midday in the summer, we also played in the rain in rainy times, but we never caught the flu [...]. (Verbal information)[15]

The question of allotments of buildings and constructions of diverse infrastructures of the city, must be made respecting some characteristic aspects of the geographic location of the city on the imaginary line of the Equator, being distinguished among these aspects the latitude and solar declination that influence the projection of shadows, useful in the planning of the forestation and thermal comfort of the city.

---

[15] Subject S.D.A., 57 years old, ordinary citizen, city of Macapá, Amapá, Brazil.

1.2 Symbolism of the Equator line in the city of Macapá

Starting with Todorov (2013), it is possible to affirm that the symbolic aspects are represented by the various forms of expression and interpretation of facts, which can be presented verbally or through everyday things and objects.

The location of Macapá is in a region of easy contact with countries of Central America, North and Europe, it transforms the city in potential of commerce and tourism, looking for products of mining, wood, livestock, pisciculture, açaí, Brazil nut, rubber, andiroba, copaíba and medicinal plants. This facilitates the dissemination of the history, culture, religiosity and tourist potential of the city, such as: the Fortress of São José do Macapá, Marco Zero, Zerão Stadium, Pan of Amapá, Sambódromo de Macapá, Pedra do Guindaste, Trapiche Eliezer Levy, Sacaca Museum, among others. Below are some testimonials from residents of this city describing the experience of living in the so-called city in the middle of the world.

> [...] we are in the centre of the world and very privileged, in a strategically geographical location for French Guiana, arousing curiosity to Brazilians from all over the country to visit here, added to the fact that we have a special sunbathing made in the middle of the world, with very fresh shades of our tourist centres [...]. (Verbal information)[16].

> [...] moments of festivities represent violent moments because the thieves know that the city is full of tourists with money and valuables, lowering this trend in the off-season tourist season [...]. (Verbal information)[17].

The recognition of the strategic position of the city, crossed by the Line of Ecuador, led the Department of Historical Heritage of the Municipal Secretariat of Culture (DPH), the body responsible for the city's nomenclature, to approve the allocation of names related to the Line of Ecuador to public institutions Photo 1.

---

[16] Subject S.M.M., 38 years old, SETUR technician, city of Macapá, Amapá, Brazil.
[17] Subject V.A.M., 30 years old, ordinary citizen, city of Macapá, Amapá, Brazil.

Photo 1 - Marco Zero Campus of Ecuador

Source: The author (2017).

The Marco Zero campus in Ecuador is the headquarters of the Federal University of Amapá (UNIFAP), which received its name based on the fact that it is located in the Marco Zero complex, crossed by the imaginary line of Ecuador. The campus consists of six (06) departments, namely: Department of Biological Sciences and Health (DCBS), Department of Exact Sciences and Technology (DCEXT), Department of Education (DEDU), Department of Arts and Letters (DEPLA), Department of Philosophy and Human Sciences (DFCH), Department of Environment and Development (DMAD).

Regarding the perception of the campus academics, it should be noted that research projects are in progress seeking to analyze the influence of the Equator line and its phenomena on the daily life of the campus and the city in general, as stated by V. A. G. (verbal information)[18]:

> [...] we are taking advantage of the strategic location of our campus in a privileged geographical area due to the passage of Ecuador, and through the GEOequinox research group, we are designing research projects on the influence of equinoxes in the city of Macapá. Similar projects are being done by the physics course, sometimes in partnership with our research group [...].

The perception of A.K.N. (verbal information)[19] is that:

---

[18]Subject V. A. G., age not indicated (I/I), SETUR, city of Macapá, Amapá, Brazil.
[19] Subject A.K.N., age not indicated (I/I), UNIFAP, city of Macapá, Amapá, Brazil.

> [...] although architectural knowledge has advanced a lot in Brazil, the
> projection of the Marco Zero Campus in Ecuador, did not observe the
> geographical location on the Equator line, applying thermal comfort criteria
> and solar diagramming, in order to design classrooms, teachers' offices, parking
> garages and circulation pavements, which end up getting sun from noon local
> [...].

Getting sun at the hottest hours of the day is what generally happens in offices, residences and car parking places throughout the city. This happens because at latitude zero degrees the sun always rises apparently in the east position and sets in the west throughout the year which means that the shadows in the period between 11:30 to about 15:30 (the hottest time of the day) the exposed sunbeds for housing and other forms of day-to-day use are arranged in the scorching sun, as demonstrated by (PHOTO 2 and 3).

Photo 2 - Shadow at the Marco Zero Campus of Ecuador Macapá

Source: The author (2017).

Source: The author (2017).

As illustrated in the photos above, the places planned to park cars and pedestrians are in the sun at the hottest time of the day, about 11 hours and 30 minutes, when the photos were taken until around 15 hours and 40 minutes, and in the same way happens in the car park where the shade of the trees project out of reach, leaving the cars in the sun, already on the treadmill exposing the pedestrians to the sun.

In order to alleviate high temperatures, residents adopt techniques such as flannels[20] that place a cardboard on the car window, according to J.V.C. (verbal information)[21].

> [...] we don't keep anyone's car, we just control the heat inside the cars, we call it shade that gives money, we have a technique to control the sun so as not to burn the inside of the car, we put the cardboard in a position that allows comfort inside the car. The priority of the flannels is not the safety of the car but to attenuate heat by shading, we do not keep the car, unless the owner asks for it, when the driver arrives, he offers from a real forward. There are days that I won in a single day seventy reais [...].

The advantage of this practice is to shade the interior of the car, avoiding heating the steering wheel and the seats, which can discourage the driver in some parts of the body. About this problem, the flannels explained that the placement of cardboard requires practical domains of urban geometry combined with the apparent mobility of the sun and the direction of shading. That is,

---

[20] Clandestine car guards in the streets of big cities.
[21] Subject J.V.C., 22 years old, ordinary citizen, city of Macapá, Amapá, Brazil.

> [...] in the streets arranged in the direction of Novo Horizonte - Zerão (North - South), we place the cardboard on the right side (East), of the front window of the car, while in the streets outlined in the direction of the Dawn Downtown district, we already know that it is necessary to place the cardboard covering two sides, the front part of the car window and the left side, because they are the places that suffer more heat from the sun. (Verbal information)[22].

Photo 4 - Shadow-papel for cars in Macapá -AM

Source: The author (2017).

The firstFoto 4 refers to the placing of cardboard for shading the interior of cars, was taken at Rua Duca Serra, another prominent place, where flanelinhas perform shade-work is on Rua Francisco Azarias Silva C. Neto in front of Casa do Artesão de Macapá,

Returning to the nomenclature of Public Institutions emphasizing the line of Ecuador and its phenomena, they are also symbolized by the sports stadium Zerão, geographically located in the city of Macapá. Its particularity lies in the fact that the midfield line, coinciding with the imaginary of Ecuador, is nicknamed Zerón, that is, the pitch has been designed in such a way that each team and the fans are some in the northern hemisphere and others in the south. The official name of the stadium is Milton de Sousa Corrêa, used for matches of Amapá teams, such as: Amapá Clube, Esporte Clube de Macapá, Trem Desportivo Clube, Santo - AP and Ypiranga Clube.

Another public element is the Marco Zero astronomical monument of Ecuador, which symbolizes the Macapaense identity, located in the square with the same name, in the neighborhood of Jardim Marco Zero. It is limited to the North: with the travessa Irineu L. de

---

[22] Subject D.J.A., 23 years old, ordinary citizen, city of Macapá, Amapá, Brazil.

Souza; to the South and East: with the highway Juscelino Kubitschek (JK) and; to the West: with the travessa Geraldino Lopes Souza, see the external leisure and parking area (PHOTO 5 and 6).

Photo 5 - External part of the Marco Zero monument of Ecuador - Macapá/AP

Source: The author (2016).

Photo 6 - External part of the Marco Zero monument of Ecuador - Macapá/AP

Source: The author (2016).

The monument was built with marble and stone, able to visualize equinox phenomena, marking the passage of the sun in the imaginary line of the equator, latitude zero (00°) degrees of the world. It has a sundial or obelisk, about thirty meters (30 m) high and represents the imaginary line twenty meters (20m) long, inaugurated in 1987. The first structure of the monument was created in 1945, with a representation only terrestrial. It is in this place where visitors come to appreciate the architecture of the astronomical infrastructure and to realise the dream of being on the Equator and in both hemispheres simultaneously (PHOTO 7).

Photo 7 - Marco Zero Astronomical Monument of Ecuador - Macapá/AP

Photo: The author (2016).

> [...] the monument was once represented by a concrete line about twenty (20) metres long in the same place where it is situated today, in a geographical area about three (3) kilometres from the city centre [...]. (Verbal information)[23].

To highlight its architecture, this monument contains a garden and is located in a roundabout with traffic flowing to different directions of the city. It has a straight line that indicates the east and west direction of latitude zero. It contains about thirty (30m) metres in height, in the lower external part where the roundabout is located, beginning with Equatorial Avenue and the others that interconnect the monument to the Marco Zero Tourist Complex of Ecuador. The upper part presents a space used for observations of the moment when the sun is on an obelisk during equinoxes, making it easier to read the sundial.

---

[23] Subject S. M. S. L., age not identified (I/I), technical, city of Macapá, Amapá, Brazil.

Inside, it has one (1) typical regional food restaurant, one (1) regional handicraft shop, dining room and external space for fraternization, also used for scientific exposition, contemplating an auditorium and handicraft shops (PHOTO 8 and 9).

Photo 8 - Internal part of the Marco Zero monument

Source: The author (2016).

Photo 9 - Internal part of the Marco Zero monument

Source: The author (2016).

Most of the handicraft objects, sold on the site, are made to imitate the Marco Zero monument or the imaginary Line of Ecuador representing the thought of people living in a city in the middle of the world.

The Amapá State Secretariat of Tourism has used the monument as a place to disseminate scientific knowledge, promoting lectures and academic fairs. Participating in the fair are teachers and students from Amapá's elementary schools and universities, especially the Federal and State public ones, including the Geoequinox group and the Mirzan astronomy club. The monument is open from 8 a.m. to 5.30 p.m. from Tuesday to Sunday, with the help of technicians from the Secretary of Tourism.

Of the attractions of the monument the egg experience stands out (PHOTO 10): "[...] the egg is in balance without falling when placed on the horizontal mast that symbolizes the imaginary line of the Equator, aligned with Equatorial Avenue that crosses the city to the Amazon River [...]. (Verbal information).[24]

Photo 10 - Egg on the mast that symbolizes the line of Ecuador

Source: The author (2015).

Balancing the egg on the horizontal mast, which symbolizes the imaginary line of the equator, is one of the curiosities that attract tourists. As the tourist X.Z.B. says (verbal information)[25]:

---

[24] Subject R.V.R.S. 35 years old, city of Macapá, Amapá, Brazil.
[25] Subject X. Z. B. 31 years old, tourist, city of Macapá, Amapá, Brazil.

> [...] I have come to fulfill my dream of being in the middle of the world, seeing the egg standing still without falling, turning on the faucet or pulling the sink and seeing water spinning in the opposite direction of the other hemisphere, in short, there are so many curiosities: like the flowers in one hemisphere, the birds travelling from one hemisphere, so many things that we only hear, or see through the posts on Facebook of friends and family [...].

One of the attractions during the equinoxes is the Fest Jeep[26] which is usually held in the car park of the tourist complex "Meio do Mundo" located between the Sambódromo, Zerão Stadium and the Marco Zero monument in Ecuador (PHOTO 11). "[...] is the track where we hold the event, it has eight hundred meters (800 m) of winding track where the competitors must cross four times the northern and southern hemisphere during the competition". (Verbal information).[27]

Photo 11 - Fest Jeep track during the September equinox

Source: The author (2016).

The event takes place since 2009 and is also known as Off Road which means: driving off road. One of the goals is to celebrate the passage of the natural phenomenon of the equinox, bringing together the public and talent. In 2015 about twenty-six (26) Jeep cars, twelve (12) bikes, four-wheel motorbikes and some people accredited to sell food, shirts, art objects and other items during competitions were registered.

---

[26] A radical competition that needs a lot of skill from the driver to know the right moment to make a maneuver, brake or step on the accelerator on the treacherous improvised track in the sambadrome car park.
[27] Subject J.M., uninformed age, city of Macapá, Amapá, Brazil.

Photo 12 - Bicycles (*bikes*) attending the Event

Source: The author (2016).

Photo 13 - Jeeps participating in the Event

Source: The author (2016).

The starting point is at the Equator line, the Jeep driver must do four laps crossing the imaginary line that divides the field from the Marco Zero Complex. The competitors make four laps in the field crossing the Equator in the South and in the North, it is also used for hiking, training dogs.

One of the aims of the *Jeep Fest* is to bring society together and respond to the positive dissemination of Amapá on an *in door lane* set up inside the park including evidence of transposition of objects against the sundial. The event is attended by Brazilians from several states and neighboring countries such as French Guiana, Suriname, Tobago and Trinidad. The event starts with a reception of male and female drivers on a track 800 meters long in a winding track with two double batteries, with the drivers racing next to each other, one outside and one inside, with a reversal in the next lap. The route is similar to the city of Brusque in Santa Catarina. For example, in the year 2016 went to the *festjeep*, pilots from Rio de Janeiro, crossing the country to Macapá, Paraíba and Santa Catarina. This competition includes, since 2015, cars known as cages and *mountain bikes*. On the first day they make a presentation for tourists and residents of the city. The competition has the qualifying phase and final that ends with the award of champions.

About this party, the social subject disagrees:

> [...] we celebrate the arrival and passage of the sun in our city, during a week, being one of observation of the phenomenon happening the scientific fair. Another competition of *bikes*, three of *Jeeps* and a football game, totaling five days of celebration. This year, the VIII Middle of the World is organized by the Club Jeep de Macapá in partnership with the French Consulate, at the opening ceremony we sang the National Anthem of France and Brazil, raising the flags of the two countries present in the event [...]. (Verbal information)[28].

To point out that there are other places in Macapá that are visited and have a daily influence on the residents who live in the city as a *place*[29] of housing and various services: Nossa Senhora de Fátima, Floriano Peixoto, Rio Branco, Chico Noé, Nossa Senhora da Conceição, Praça da Bandeira, Complexo turístico e de Lazer da bordla do Macapá. Included in the list are all cultural and landscape heritage consisting of the Fortress and Church of São José de Macapá, the Joaquim Caetano da Silva Historical Museum, the AOB Headquarters, the Municipal Market, the Bacabeiras Theatre, the Black Culture Centre, the Sacaca Museum, the Ribeirinho Fair, the Indian Lagoon, the Mato Well, Quilombos do Curiaú, Archaeological site of Curiaú, archaeological site of Pacoval, archaeological site of UNIFAP, archaeological site of Fátima do Maruanum and the archaeological site of Ambé, including the cemeteries of Nossa Senhora de Conceição, Coração, Maruanum, Curiaú, Pedreira, São Joaquim do Pacuí, Santa Luzia do Pacuí, Carapanatuba and do Bailique.

The superimposition of shadows as a phenomenon that elucidates the equinox can be observed from anywhere in the city of Macapá, leaving for the Landmark Zero monument the observation of the moment when the sun passes through the obelisk, symbolizing the exit and arrival from one hemisphere to another. What is interesting is that in the city most residents know that the overlapping shadows are one of the indicators that the sun is closer to the equator, at the same instant that it forms a right angle, announcing the equinoxes. This phenomenon can be

---

[28] Subject M.M., of uninformed age, organization of *Festjeep*, city of Macapá, Amapá, Brazil.
[29] Place or geographic physical space.

observed at any point in Macapá, and it is only possible to move to mark zero, the resident who needs to see the sun through the obelisk and follow the programming of the ritual of arrival and passage of the sun through the city.

On the equinoxes and shadow overlap, S.D.M. and H.L. residents say:

> [...] my family and I noticed for the first time that the shadows overlapped in Macapá, when we were at my father's funeral on September 22, 2013. We were all frightened when in that very hot sun, my niece noticed that nobody had a shadow, so we thought that it had to do with the spirit of the deceased. But we learned later that many people had missed the shadows that week that had disappeared, there we were relieved [...]. (Verbal information)[30].

> [...] I understood that at the time of the equinoxes, the benches in the square were not shaded. That all the shadows you need to have a picnic and lunch with family and friends, are far from the benches. The only way is to spread a cloth, a cardboard or to sit on the ground, but very close to the stem of the tree, where there is shade for us to hide from the sun. One day, my colleagues and I raised a discussion about the possibility of resizing the benches, closer to the stem [...]. (Verbal information)[31].

The lack of shadows on the banks of public squares at the hottest hour of the day forces most of the city's population to frequent these places at night. As the following statement goes:

> [...] in this city there is a lot of *gay*[32]movement, which I am part of, among several objectives, the idea is to accustom the homophobes of[33] the city on the issue of homosexuality; for this we need to be visible; but as our squares do not have enough shade to house everyone, we enjoy the squares at night, meeting almost every night, during the day it is not convenient, there is too much sun, there is no way to stay in the square in the afternoon without shadows [...]. (Verbal information)[34].

The perception about the projection of shadows in the city has been a debate even in the public policies of the municipality, as the speech states:

> [...] the subject of shadows is complicated in our city, I can comment that a project of the Municipal Department of Urban Development and Housing (SEMDUH) was in progress, aiming to standardize *public stalls, called* 'Calçada Livre' of the streets and avenues of the city of Macapá, making Macapá more accessible to people. Despite efforts, the banks of the public squares are abandoned, there are even pedestrians walking outside the pavements, because the shadows are still projected in the middle of [...] Street. (Verbal information)[35].

In addition to the nomenclature of public institutions, there is also a tendency to name

---

[30] Subject S.D.M., 43 years old, city of Macapá, Amapá, Brazil.
[31] Subject H.L., uninformed age, city of Macapá, Amapá, Brazil.
[32] An English word, generally used to describe a homosexual man or woman.
[33] Prejudice of some people or groups against homosexuals, lesbians, bisexuals and transsexuals.
[34] Subject G.M., 22 years old, ordinary citizen, city of Macapá, Amapá, Brazil.
[35] Subject S. M. S. L., of uninformed age, SETUR, city of Macapá, Amapá, Brazil.

streets, avenues, neighborhoods, hotels, commerce, workshops, public transportation, etc., using the name of the equator line and its phenomena. This shows how much the residents of the city make of this line their identity and pride to live in a place marked by its geographical particularity of being at latitude zero, full of phenomena typical of its exclusive location.

The perception of the importance of this geographical location of the city on the equator and under the influence of its phenomena is not only registered by the nomenclature of public and private institutions, but also by the matters corresponding to daily life, as a social subject says about the apparent mobility of sun and shadows.

> [...] I noticed that at home that the sun changes position, when I paid attention to where it rises in relation to the window, I also paid a lot of attention to the shadows from there and I noticed as soon as they change position every day and every hour, it was easy to notice when tying the hammock. I even know how to use the shadow to control the time, my father taught us, because at that time, there were no wall clocks in our house, nor that particular one for each brother, nobody had [...]. (Verbal information)[36].

In addition to the timetable, city residents have reported that they determine the equinox seasons through daily senaries, and the entire first half of each year, many neighborhoods in the city are flooded, this always happens between the months of March and April. It even makes it difficult for aeroplanes that go to Belém to land because they cannot because of the heavy rain. During this period, the floodplains are planted in the floodplains, pedestrians sell many umbrellas, and residents of the hangover go to live in the high zone. The beaches are left without people because they all disappear, people are left without to buy the sales there on the beach. The equinox of the droughts is the time of the beaches, the fruits and vegetables are abundant at low prices, the air conditioning works non-stop everywhere, elevating to high energy payments; the institutions of events and leisure earn a lot in the second semester, all sell a lot, from the small trader.

Despite knowledge about the impacts of the equator line, most residents do not go to the landmark monument for observations of the moment the sun passes through the obelisk, as one of the social subjects of the research explains:

> [...] visiting the landmark is of no interest to us in the city, I always came here with friends just to enjoy the movement of people, we never saw anything I, my friends. These guys who stay here only explain things they have on the internet, like what equinox is, where it passes, but they never showed the phenomenon itself happening, I don't even know how to see it in this monument [...]. (Verbal information)[37].

Adding on the same issue, other residents and tourists say that they only go to meet new people, to appreciate tourists and girls of the city because the movement in these days is great, but

---

[36] Subject F.L.S., 59 years old, ordinary citizen, city of Macapá, Amapá, Brazil.
[37] Subject P.M.T., 20 years old, ordinary citizen, city of Macapá, Amapá, Brazil.

it is never for contemplation of the equinox phenomenon, claiming that they can visualize the most important moment in their homes, which is the overlapping of shadows lasting at least a week, in that order of ideas, going to the landmark monument zero see the sun, is a tourist thing.

One of the reasons for the lack of motivation is the difficulty in contemplating the moment when the sun is over the equator, as one of the residents said:

> [...] I was able to see the sun in the hole this year because a UNIFAP teacher taught me how to see, but many return without having seen it, which is of no interest to come anymore, giving propaganda that it is not possible to see the sun passing through the other side of the hemisphere [...]. (Verbal information)[38].

The lack of contemplative habits leads residents, including some UNIFAP students, to think that the performance of the equatorial line and its phenomena have their localized impacts only on the equatorial avenue and surroundings. In view of this, a work of questioning started in two thousand and eleven (2011) regarding perception is being done with the students of the first period of the Geography course. The results show that most of these students, who do not know and have never visited Marco Zero, do not know about equinoxes, the equator line and its importance, especially for the city of Macapá. If the students of the geography course, where these programmatic contents are part of the teaching curriculum, do not know enough about this subject, imagine the ordinary or unstudied inhabitants of the city. The measure found to minimize this problem was the creation of the GeoEquinox nucleus to produce multipliers of this knowledge in the University and society in general.

Although city dwellers use the equator and its phenomena in their daily lives, there is still a fragility about the theoretical basis and its relationship with practice. This shows more and more how the new generation of city dwellers is distancing themselves from nature, differently from the inhabitants of the village of Aqui in Mozambique that there is a continuous and active interaction, producing daily models of perception and survival based on this analogical relationship.

The subjects questioned in Macapá about the size of Ecuador and its impact, most of them answered that the imaginary line of Ecuador started at the roundabout of the equatorial complex and ends at the equatorial avenue where it disappears, nor reaches the Amazon River. And that the sun and the shadows begin their gymkhanas there, already in the centre or in the river it was not possible to contemplate any phenomenon related to the passage of the equator. For this group, the line is very short, that is why they do not feel or observe anything related to its effects in houses and they have never seen anything of what others can see on the landmark monument zero.

Despite this observation, they said that line phenomena are used to determine the seasons, and that during the March or Autumn equinoxes, the neighborhoods located in the northern part of

---

[38] Subject B.C.L., 43 years old, ordinary citizen, city of Macapá, Amapá, Brazil.

the hemisphere, their trees begin to blossom, while those on the southern side of the city, the trees fall their leaves, the opposite phenomenon occurring during the September or spring equinoxes.

The testimonies show that although the city dwellers have perception of the value of the imaginary line and its phenomena, most of them do not have a mastery of basic concepts about the extension and dimension of the line in the city and the world, even thinking that it belongs only to the city, its impacts being limited to the Marco Zero tourist complex and Equatorial Avenue. One of the impacts referenced by the residents is the shadow as an indicator of the moments of equinoxes, used to demarcate the seasons, as it happens with the solstices for the inhabitants of the village of Aqui.

1.3   Haqui village - Mozambique

1.3.1   Physico-geographical aspects

In its apparent movement, the sun returns to the north when it reaches its maximum mobility to the south over the imaginary line of the Tropic of Capricorn (T.CAP), at the instant that its rays are focusing on a slope of forty-six degrees and fifty-four minutes (46°54') over the village of Aqui and other places crossed by this line.

Geographically, the village is located in Inhambane province, Massinga district in Mozambique (MAPA 3). To the south it is bounded by the village of Licunha, to the north by Malova, to the east by the village of Chissindane and to the west by Cangela, in an area of approximately forty square kilometres (40 km²) of territorial extension. The seat of the town is located at the astronomical coordinates, twenty-three degrees, twenty-six minutes, twenty-two seconds (23°26'22" S) South and thirty-five degrees, fifty-two minutes, twenty-two point two seconds (35°52'22,2" E) East.

Combining data from the Strategic Plan for District Development (PEDD, 2008-2012), with those from the National Institute of Statistics (INE, 2010 - 2012) and Farré (2012), it is possible to say that the village is located in the most populated district of Inhambane province, with a population of over 250,000 inhabitants. The district is divided into two administrative posts: Massinga headquarters and MassingaChicomo, with five locations: Rovene, Guma, Malamba, Lionzuane and Chicomo.

The village of Aqui belongs to the locality of Rovene and is run by a second level community leader[39]assisted by a consultative assembly made up of a number of land or Madodas chiefs, who assist in decision-making as well as in the dissemination of values, ethics and morals

---

[39] It is one that does not ascend by heredity or inbreeding ties, but has a strong influence on decision making in the daily dynamics of the village.

in the village. The indigenous population is of Ba tswae ethnicity who speak the Xitswa language[40], Portuguese is learned and spoken normally by those who have attended school.

Map 3 - Geographic location of the Povoado de Aqui - Mozambique

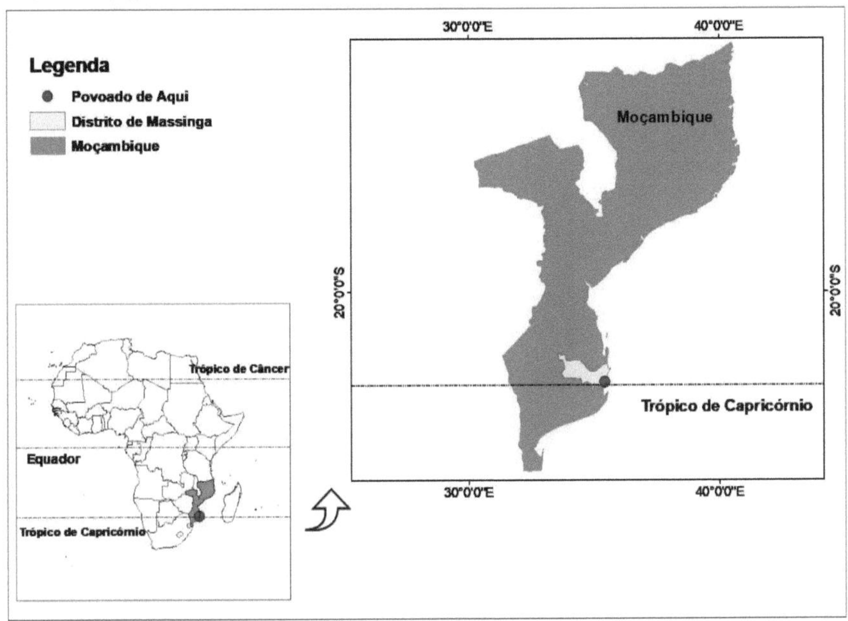

Source: The author (2018).
Note: Using HQIS 2.18.15 *las palmas*.

In order to evaluate the seasonality of the climatic elements and to establish the averages of standard deviations for all months, showing the behaviour that differentiates the dry season from the rainy season and the distribution of rainfall values per quarter, it was deduced from the methods of climate classification of Koppen (1928), Thorntwaite (1948), Strahler (1978) and the Mozambique climate map of the National Institute of Statistics (INE, 2012), to substantiate that the village of Aqui has a warm tropical climate, with a high level of humidity in the east, which changes as it walks westwards, becoming dry without great variations.

The average temperature is 22.9°c, the absolute maximum is 35.9°c and the absolute minimum is 19.2°c, with a relative humidity of 79.8%, corresponding to a precipitation of 55.8mm. The months of November to March are rainy, with rainfall of more than 1054 mm; July, August, September and October are dry with rainfall of less than 30 mm. April, May, June, and

---

[40]It is part of the Tonga (changana, ronga and xitsuaoutsua) branch of southern Mozambique.

July is the period of low temperatures, placing the village in group A of the Koppen classification,[41]and W of Thornthwaite[42](CHART 4).

Chart 4 Thermopluviometrics

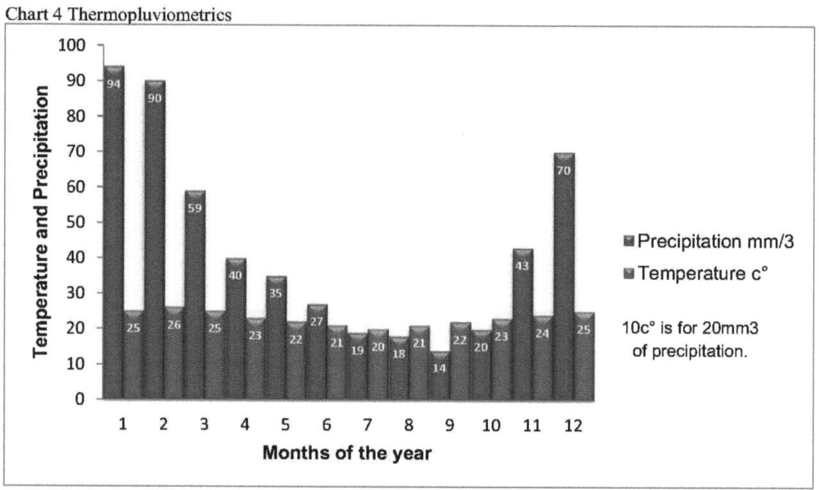

Source: The author (2018).
Note: Using data from the Statistical Yearbook of Inhambane Province published by (INE 2016).

The high precipitation values are registered in the months of December, January and February when compared to the driest month there is a rainfall difference of 157 mm.

Here it is not crossed by any river, taking advantage of the watersheds of other villages such as Mahocha and Malova which are streams; there is a perennial lagoon here43 which plays an important role for seasonal fishing and irrigation of fields for family agriculture practised on sandy and clayey limestone soils [...]" (ALMEIDA, 1959, p. 31), mainly in the dry season producing legumes, cane, rice, cassava, bananas, vegetables and various tubers. The soils are black and clay-coloured, fertile and of great productive capacity, with the capacity to conserve humidity, from the banks of rivers,[44]classified locally as male or hydromorphic, Muchangos (1998); Ombe (2007), and Tchernozen, according to the Russian Dokuchaev, developing under the influence of a water table, sometimes presenting the façades of water-saturated areas. These are soils with high production capacity due to water retention over a long period of time.

---

[41] Based on the interaction of rainfall and temperature, the village of Aqui is part of the group of rainy tropical regions. Because it has an average annual temperature "[...] of over 18°C and annual rainfall of over 750 mm". (ALMEIDA, 1959, p. 31).
[42] Classified from the atmospheric humidity resulting from the evapotranspiration of living beings and soil, placing the village of Aqui in the group of Aw (tropical humid) climate regions and as we move towards the osemiarid interior.
[43] Because it only appears in the rainy season.
[44]Black clayey soils, fertile on the banks of rivers, used for agriculture in Mozambique.

The ground cover has undergone conversion, being formed by shrubs that gather in isolated bushes, reminiscent of a forest of miombos, which has been replaced over time by human activity. The landscape is alternating with the coconut plantations, which constitute the main visual horizon and wealth of the local populations. Although there is no river here, the residents take advantage of the Chicamba, Guizugo and Mahocha river basin, situated in neighbouring villages, to exploit resources, practising family farming, vegetal and animal extractivism, removing clay for maticing or towing houses and making clay pots, bathing, washing clothes and various materials, etc.

Based on Almeida (1959), Muchangos (1998), one can say that the orography of the village of Aqui is plain, with points not exceeding two hundred (200 m) metres in altitude. The study region belongs to the recent cretaceous as the predominant system, being of the lacustrine quaternary, with marine limestone in small spots of sedimentary rock formations, volcanic and metamorphic complexes, organized in sand of coastal dune and sand in the region near the beach of Chiduca while to the west, there are rocks of the superior pre-Cambrian that allow the development of shrub and arboreal vegetation.

The soils on the border with Chiduca are whitish sandy with high water permeability or difficult to retain water and humus, so less fertile for agricultural activity and susceptible to erosion, especially on the slopes. In the central and interior part they are sandy-clay type, which together with a tropical climate, in the rainy season the conditions become optimal for the practice of family agriculture, allowing the production of crops such as maize, butter beans, nhemba, yoke peanuts. Forest species of economic value (Messassa and chanfuta) have become extinct due to human exploitation in the past, as the following social subject points out:

> [...] many species of plants that formed the vegetation of our village have disappeared, even animals such as elephants, boars, *llangos* were around here, my mother tells us, but they have disappeared; but some still keep dodging man around here, such as rabbits, *vondos*[45], monkeys, partridges,[46]which resist in the small forests down the coast [...]. (Verbal information)[47].

Despite the extinction of several species of local flora and fauna, there are still resources of common use that are considered the wealth of the village, such as bat guano48, limestone, the lagoon, flora, fauna, the Mahocha stream and Malova, which even being less voluminous constitutes the main hydrographic basin that irrigates the agricultural fields and areas of pasture and water source for domestic use. Already in the dry period the resources of the forest are used, as is the case of the following images (PHOTO 14 and 15).

---

[45]Known as capybara in Brazil, whose scientific name is *hydrochoerushydrochaeris*.

[46] A galiform bird belonging to the Phasianidae family, whose scientific name is *rhinchoutusrufescens*.

[47] Subject J.A.C., 48 years old, village of Aqui in Massinga, Inhambane, Mozambique.

[48]Bats' faeces accumulated in caves and fronted trees, generally used to fertilize the soils, due to their high levels of nitrogen, making them more productive with the addition of faeces.

Source: The author (2015).

This plant is used for food in the dry season, its leaves are used to make craft objects such as baskets and wallets, ropes, among others. The fruits are made from various juices, mousse, etc. Another wild plant that serves as a survival tool in times of drought is the macuacua, scientifically known as *Strychnosmadagascariensis*.

Source: The author (2015).

The macuacua is a fruit native to regions with low humidity, often found in savannas of the national coast, and that coincides the ripening of its fruits with the dry season, and of easy conservation, from it are manufactured several food derivatives such as molasses, oil, flours that serve as alternative food, vitamins and minerals much consumed in the dry season by the inhabitants of the village.

Socio-economic aspects

Outside the exploitation of the flora, the village practices other activities of the primary sector (agriculture, family livestock, hunting, fishing and vegetal extraction), handicraft and tourism in the preliminary phase encouraged by the tourist authorities located on the coast. In addition to beach tourism, various types of fish are taken advantage of, as well as sea exports.

Agriculture is familiar and itinerant[49], the main crop species are cashew (*Anacardiumoccidentale*), mafurreira (*Trichilia emética*), mango (*Mangifera indica*), citrus (*Strichnos spp.*), coconut (*Cocos nucifera*), maize (*ZeaMays*), rice (*Oryza sativa*), groundnut (*Arachishypogaea*), beans (*Phaseolusvulgaris*), banana (*Musa spp.*), sweet potato (*Ipamoea batatas*), vegetables, sugarcane (*Saccharumofficinarum*), cassava (*Manihotesculenta*). Rice and sugarcane are generally grown on the margins of males[50]. Production methods are domestic using burnt and short rope hoe, animal traction, land fallow system using large areas and polyculture. A research subject's statement on this subject follows:

> [...] our unplanted priorities are coconut, cashew, mango and citrus, these are the most cultivated species. Now the coconut tree covers our entire village. We use all its components of the coconut tree, for example, in medicine the root is used for the treatment of various diseases, including toothache. The trunk is used for the construction of our houses, it is made of clay, laths, wood and the *macute*[51] is used to cover *shacks*[52]. Children use the nerves of the leaves to make toys such as vignettes, flutes, baskets, among others. The branches and the stem serve as firewood for cooking, to manufacture brooms and brushes. The coconut fruit is used to season broths, to produce oil, cake and the pomace for animal feed. The thatch sap produces a drink known as *sura*[53], coconut we sell by units or in *copra*[54] form this exercise also happens with mafureira and cashew [...]. (Verbal information)[55].

---

[49]It uses traditional land clearing techniques, such as: burning and rudimentary instruments.

[50] Black clay soils on the banks of rivers.

[51] Coconut leaves.

[52] Locally used to refer to housing built from plant material, the term has evolved to include the conjugation of uses, with tents being built from local material: zinc and cement. The tent is usually rectangular in shape.

[53] Delicious juice made from coconut. When sweet it can be used for fermentation of wheat for cake making. The sura is sold as an alcoholic beverage and left for a long time in the container turning into vinegar used in food and for wound disinfection.

[54] When peeled, broken and left to dry in the sun, the coconut turns into copra, see Photo 48, which can be sold in factories for vegetable oil and soap production. The leftovers are used as pomace for animal feed.

[55] Subject H.F.C. 56 years old, village of Aqui in Massinga, Inhambane, Mozambique.

Among the various ways of using the coconut tree, one of them is to peel the coconut in order to sell or to break/break it in order to sell it in copra form. Still on the food and customs of the village of Aqui, we have to point out that when it rains a lot, edible mushrooms (*calangane, matlimbe, rundzie, finhane, Maotchane*)[56] appear around in the bush, which can be collected, cooked with water and salt, made with peanuts, of coconut. They can be dried in the sun or smoked in the barn and preserved for dry seasons. There are poisonous mushrooms one of which is called Nhatchecuane in xitsua.

In the village there are activities that are usually women's and others for men and children. Often the construction of tents, bathrooms[57], fishing, hunting and raising the male child are done by the father while cooking, washing and teaching the daughter how to behave towards society and in her future home is done by the mother. Apesca is not just anyone who makes a few specialized men for activity, who bring from the sea the species Xerewa, the red fish, fish, shrimp, octopus, crab, squid, amejoa, garopa, thieving fish, stone fish. Some women just buy and resell the fish.

The most domesticated animals are: the ox, the donkey, the goat, the chicken, the duck, the bush hen, the goose and the turkey, the cat and the dog. Several types of birds, rats, squirrels, rabbits and *roundworms* are generally hunted.

The population density of the village is three thousand and ninety-two (3,092 inhab.) inhabitants living in (773), families led by a first class community leader[58]assisted by the Madodas[59], who are a species of adult people, trusted by the leader and legitimated as suitable by the inhabitants of the community.

The role of leader was shaped by Decree No. 15/2000 of 20 June, assigning positions to Mozambican citizens who show aptitude and confidence to take responsibility for their village or community. In other words, to be a leader, the individual needs to deserve trust and credibility from the members of the village, he or she must be a person who transmits certainty in the hearts that will help to solve and bring to government all the problems of the village. But it must also be someone who has respect, knows well and respects traditions, that is why at the time of the survey they accepted Alberto Covele as their leader. The chosen one has no salary for the position, but must commit himself as if he were a job. He must mobilize and participate in the search for solutions to various problems in the village, such as: pregnancy and premature marriages,

---

[56] Names of mushroom species in *Xitsua* language.
[57] Bathroom.
[58] According to the Pedd (2008-2014), it is the one that does not ascend to power through ties of heredity or inbreeding, is recognized by the residents as authority, exerting greater influence on decision making for the daily life of the village.
[59] Elderly people of recognised experience and authority are respected and accredited in the community for their exemplary way of life.

adultery, Levirato or *Kutchinga*[60], land conflicts, water plumbing, installation of electricity in the village, etc.

This town does not have a primary school of first and second[61]grade, the Escola Secundária Maria de Luz Guebuza de Mupakulane teaches first cycle, that is, from eighth to tenth grade. The classrooms are made of masonry, built of stone and zinc slabs, others are made of zinc, both on the walls and on the upper roof.

About the physical condition of the school one of the social subjects said that:

> [...] my son complains of headaches in hot weather, the zinc room is very hot just in this heat of Massinga, for a child of thirteen (13) years, locked in a place full of others touching sweat is sad, I can even say that it is worth the other my daughter who studies under the shade, there in the other school, it is preferable even [...]. (Verbal information)[62].

The school initiation corresponding to grades 1 to 7 does not take place within the village, forcing the children to travel long distances to neighbouring villages.

In addition to failures in education with a shortage of primary schools, forcing children of low age to walk long distances in order to have access to the first grades, the village is also abdicated from health offered by the government in a public and accessible manner.

Regarding basic sanitation and water supply, electric power, etc. ...] here in our village, our water supply is damaged, but, thank God, we were able to get electricity and piped water" (verbal information).[63]Therefore, some symbolic elements related to experiences lived from the shadows in the village are presented.

1.3.2 Symbolic aspects of the village of Aqui

Unlike Macapá, the village of Aqui presents a spatial organization that does not obey the plans of scientific details, that is, the village has no zoning plan, standing out for landscapes resulting from the conversion of the soil, for the construction of residences and the practice of farming, predominating palm trees that configure a typical landscape of the Inhambane coast and the village of Aqui.

It is in this landscape where the experience of agrometeorology, photoperiodism, perpendicularism, thermal comfort, among others, are used for the practice of family agriculture,

---

[60]A habit or custom of the *Ba stwa* ethnic group and others of patrilineal lineage which obliges the brother of the deceased to marry his brother's widow when he leaves no male offspring, the son of this marriage being considered a descendant of the deceased. This custom is also mentioned in the Old Testament as one of the laws of Moses.

[61] Who teaches from first to seventh grade.

[62] Subject G.J.F., 37 years old, village of Aqui in Massinga, Inhambane, Mozambique.

[63]Subject S.F.M., 49 years old, village of Aqui in Massinga, Inhambane, Mozambique.

construction of residence, animal breeding, determination of time, geographical orientation among other activities of daily life.

From Chang (1974), Goyal and Builes (2015?) it can be affirmed that the inhabitants of the village of Aqui, by using the shadows to establish periods of exploitation of plants and animals during the annual cycles, are exercising the agrometeorology. The photoperiodism is established from the control of the behavior of plants and animals, which allows them to establish determined times for the use and enjoyment of available resources, practices that require experimentation on the variations of plant and animal behavior, as presented in the following statements:

> [...] the planting of agricultural species obeys the three phases of shade and sun, and there are times, for example, for lettuce, onion, tomatoes which have sometimes been in hot weather under a tree with a large shade [...][64].

> [...] cashew trees, orange trees, tangerine trees, strawberry trees and other fruit trees including those in the bush, only produce in the sunny weather mixed with lots of rain and very short nights, that is, what happens in December, so this month goes well with lots of fruit; December happens a lot because it has lots of sun, but it has good shades because the trees have good leaves, so it's a good time to gather the family under the tree, eat and talk while we drink everything there is. People grow belly and get fat in that moment [...]. (Verbal information)[65].

In the central part of the village, along the National Highway Number One (EN1), there are signs that symbolize the passage of the imaginary line of the Tropic of Capricorn through the village of Aqui in Mozambique (PHOTO 16).

---

[64] Subject S.F.M., 49 years old, village of Aqui in Massinga, Inhambane, Mozambique.
[65] Subject L.A.M., 43 years old, city of Macapá, Amapá, Brazil.

Source: The author (2015).

The concept of the imaginary Tropic of Capricorn is not known by the residents, understanding that it is an academic concept and the majority of the population of the village of Aqui cannot read, write or speak the Portuguese language because they did not go through school education, as A says. L. M:

> [...] we have seen white people parking their cars and appliances there, sometimes taking pictures and leaving. What they do with it we do not understand, but we know that the shadows when December arrives are aligned with their owner and of the trees with the trees themselves, this happens there for the end of December, this we always knew [...]. (Verbal information)[66].

Shadows here mean a lot of things in our life, when a person is born he gains shadow, so the shadow is a person's life and when the person dies he goes with his shadow. If you are an evil person your shadow will create terrors on earth in the form of a ghost. If the person was a good person, he will only appear if his shadow is called through a traditional *mhambaou* mass of fraternization with the deceased of the family. In this ceremony the souls or shadows of the deceased of the family are invoked to talk, dance, sing and help to solve the problems of the relatives who are still alive. Because of the importance of the life of the deceased for the living it is important to provide comfort for them, that is to say, the dead should be buried in a shade and

---

[66] Subject A.L.M., 46 years old, village of Aqui in Massinga, Inhambane, Mozambique.

cleansing should be done to avoid anger (PHOTO 17).

Photo 17- Cemetery shadows

Source: The author (2016).

One of the attitudes that calls attention is that the houses or residences for the living, are of simple construction and the tombs are of stoned granite, reinforcing the question that life beyond death, continues and must be well lived. Reinforcing a social hierarchy, where families with greater purchasing power provide better tombs at the disadvantage of poor families who often have neither a coffin, burying their deceased relatives without cover.

The cleaning of graves and offerings to the deceased is for some residents a practice that demonstrates their connection and conviviality with ancestry, and can contribute to the good treatment of these places to provide a joyful life without calamities and other ecological events harmful to the living.

Outside the use of shadows to negotiate good life for the deceased to the living, there is in the village experiences of time determination for shadows according to speech:

> [...] no one in the village, has a lot of money to buy a watch even for children, we always teach them how to go to school from the shade of a cashew tree or any other tree. I even know where the sea is when I am lost in the bush, just look at the height of the sun and see the position of the shade, from there I go towards the sea, there it is easy to know how to get home. That many people do when they are lost, many hunters [...]. (Verbal information)[67].

---

[67] Subject V.M.J., 50 years old, village of Aqui in Massinga, Inhambane, Mozambique.

The placement demonstrates that the elements of orientation and time counting in the village of Aqui, are framed from practices related to shadows, for example:

> [...] I can say that in the morning the shadows are fulfilled and point to the *mupelagambo*[68] or West, about noon, it is easy to get lost in the forest because the shade is usually short and very close together on the tree, but at the end of the day we know that the shade will indicate the direction of the sea or Indian ocean [...]. (Verbal information)[69].

> [...] the East point and its shadows, is the place where the sun wakes up, is the moment to wake up, always symbolizing joy; when the sun is high the afternoon has come which represents the moment of the fight for life and stress; the end of the day is the time when people are at the peak of work and its production, want to rest and resume energy in the dark of the night [...]. (Verbal information)[70].

Starting from Casati (2001), we can say that the experiences of shadows in the village are not only useful for counting time and geographical orientation, but also integrate elements that denote shadows as bad or negative phenomena.

For the villagers here, when the shadow is pointing in the westward direction or *mupelagambo is* representing the continuity of life, while this or *utchene*[71] symbolizes the end of the day and expectation to sleep and wake up alive, the night represents the greatest shadow of all, the night is the moment of darkness or of the great shadow, it also represents m*ourning*, death, the world of the dead, which is why the dead and their tombs are oriented westward. The west side and its shadows are the place of the dead and of the ancestors who incarnated the world of the gods, being able to decide on the lives of their relatives who are still alive. The shadow of the end of the day symbolises rest, which may be eternal, but also represents the wisdom of the ancestor or the beginning of the hour when the older people are able to be heard by the younger ones, telling the stories of the family and environment of the village, explaining, for example, to the younger ones about spatial dynamics and conflicts between families.

The observation of solstice shadows as a tourist or leisure activity is not yet common in the village. As described below:

> [...] but in fact since they put those plates there on the road, some gentlemen have begun to appear who stop and ask us about the shadows of the month of December because they are important for tourists. I myself have begun to pay attention, I have even seen that in addition to white people there are black people who stop there also taking pictures. Until I took a picture last December when the shadows were next to the coconut tree trunk, down by the beach [...]. (Verbal information)[72].

---

[68] Sunset, in *Ba tswa* language.
[69] Subject J.A.C., 48 years old, village of Aqui in Massinga, Inhambane, Mozambique.
[70] Subject V.J.T. 31 years old, village of Aqui in Massinga, Inhambane, Mozambique
[71] Solar spring in *xitswa.*
[72] Subject V.J.T. 31 years old, village of Aqui in Massinga, Inhambane, Mozambique.

> [...] for us, December is a time when it rains a lot, we plant, but we also pick cashew nuts and *mafura*[73]; there are plenty of vegetables, juices or juices prepared from cashew nuts or prepared from wild fruits, such as: *mambombo*[74], *massala*[75], even some even take advantage of the few left-handers that you have around here to make the *canhu*[76][...]. (Verbal information)[77].

The inhabitants of this village are aware of the fruiting of some trees from the solstice period, as described in the statement above. Below are the photos (58 to 60) with fruit from this period. In Photo 44 we have mambombo, while in Photo 59 the canhú, in Photo 60 the larger fruits are massala (a kind of edible gourd), behind the massala there[78]is passion fruit and mafura[79].

In addition to the wild fruits already presented above, mambombo is also part of this diet. It is usually taken from the forest and helps to diversify the diet of the residents.

The kennel, the massala are some of the fruits that appear in the season of food abundance, but due to their nutritional value and the flavor of their derivatives some residents have taken advantage of consuming them even though there are alternatives in the December solstice season.

For the Aquians, to read shadows is to know how to choose between cashew tree, Mafureira,[80]Massaleira,[81]coconut trees[82], mango[83] tree which produces a good tree for home, cemetery; for example in the cemetery we use *chiauau* for shade, but it also serves to signal places where a person has been buried, so the tomb will never disappear even if the relatives do not take care, the plant will remain. When someone does not have a shade tree he can improvise his hut; for example, someone has just built a new house, a *woman,*[84]and on his land there is no tree for shade, we build a barn with a lower part for shade or make flaps fulfilled in the hut. The barn has multiple uses, being for shade during the sun, normally used to store crop produce and another in winter to access fire to warm up from the cold and used as a shelter to protect oneself from the cold85 or not to get serene in winter.

---

[73] Fruit of a tree called mafureira, typical of tropical Africa and abundant in Mozambique. Consumed after softening in water for 1h or longer, depending on the thermal state of the water or the atmospheric state of the environment of the place.

[74] First photo of red fruits in *xitswa* tongue.

[75]*Strychnosspinosa*: scientific name.

[76]*Scleocaryacaffra* its scientific name. It is the fruit of a tree called left-handed in southern Mozambique, also known as canhú or amarula, used to make juices or fermented drinks, much appreciated in southern Mozambique.

[77] Subject A.L.R. 41 years old, village of Aqui in Massinga, Inhambane, Mozambique.

[78]*Strchnusspinosa* - scientific name

[79]*Trichiliaemetic* - scientific name

[80]*Trichiliaemetic*- scientific name

[81]*Strychnosspinos* - scientific name

[82]*Cocos nucifera*- scientific name

[83]*Mangifera indica* - scientific name

[84]New residence on land that still lacks shade trees.

[85] Dense, serene fog that forms in the morning in Mozambique.

Another object used for shadow projection are the haystacks, which are circular constructions, with a roof or roof formed from a hat, used as residences, but taking advantage of their shade to ventilate the environment.

Returning to the question of shape and arrangement of the shade, one can add that shadows change position throughout the day and the year, so residents pay attention before doing anything, try to analyze well whether or not the sunshine will arrive there throughout the day, this is also analyzed when it comes to tying animals (goat, ox, pig, etc.), (PHOTO 18).

Photo 18 - Using shadows to tie up pigs

Source: The author (2015).

Animal has been tied up in a position that is no longer catching shadow, and needs to be switched elsewhere. Although the coconut tree has a partial and fine shade, the image shows that the animals try at all costs to put themselves in places where the shade appears to enjoy the thermal comfort while feeding on herbs, which requires that the owner is always around to exchange the animal, as stated in the following statement.

> [...] another thing you need to know is that this thing of changing place the shade, makes the same animal tied in the shade, the owner needs to be around to change position the animal throughout the day, this so that the animal does not suffer from the sun; it happens sometimes around the tree the animal until the end of the year complete an entire turn in the tree just changing position just looking for shade [...]. (Verbal information)[86].

---

[86] Subject M.Z.P., 41 years old, village of Aqui in Massinga, Inhambane, Mozambique.

In this perspective, to turn the tree around the whole day is to assume that the shadows throughout the day change position and when the extreme points of the morning and afternoon shade are united they can form a circle, that is, by changing position the animal can have made a circle around the tree until the end of the day.

The question of changing one's position in relation to the shadow is only practical when it comes to objects and things that are flexible or easy to move from hour to hour, as the resident says,

> [...] when you store things that are difficult to change all the time, like a pot to cool water through the shade, there you already need to know the position of the shade when the sun is very hot. Ah! there yes, because if you put it wrong, the shadow disappears and the pot of water stays in the sun and everyone will suffer from thirst [...]. (Verbal information)[87].

To make the water cold it has been practical to place a pot over a tree shade, so you need to have domains of the shade trajectory throughout the day, especially at the hottest time of the day, preventing the water from being hit by the sun's heat and becoming hot without conditions to drink, according to the statement:

> [...] the first thing a person should know is, to identify trees that have good shadows, to put things in. Long, thin trees are not ideal for shadows because they have giant shadows that can always be out of reach of the things we want to protect from the sun [...]. (Verbal information)[88].

> [...] this thing of knowing how to choose the best shade tree my mother always said not to plant coconut tree inside the house because it is long and has no good shade, but to reduce its height, it should plant on its knees or belly, and the best shade is of trees with open branches [...]. (Verbal information)[89].

A technique to avoid failure in the supply of shadows to cool drinking water, is to place the container between two large objects that project shadows, so as to always take advantage of one of the two obstacles of the sun. Allied to the shade, it is essential to use clay, wood and never metal material because it is a conductor of heat, which can increase or accelerate the heating of water.

Still on the top of the trees stand out those with a conical shape facing upwards as illustrated in the image below, which have clear and leafy shadows to house more people. These types of shade are generally used to bring the family together, serving as a central place of the house, a kind of family room, to receive visits from friends and family for leisure or gatherings (PHOTO 19).

---

[87] Subject A.G.M., 43 years old, village of Aqui in Massinga, Inhambane, Mozambique.
[88] Subject A.H.T. 67 years old, village of Aqui in Massinga, Inhambane, Mozambique.
[89] Subject H.F.C. 56 years old, village of Aqui in Massinga, Inhambane, Mozambique.

Source: The author (2016).

In addition to barns and fire, used for shade and heating in the cold period, broad canopy trees are used as places for cooling but also for the production of alternative foods. In winter the trees that cast good shade during the summer spend the cold season, producing larvae of leaves known as *matamane,* known to those who have studied as a miner *fundeciturus* is well edible for the residents and nutritious. In the cold season the shade tree is also a place of fire to warm up with cold and apples, peanuts, rats, which are hunted at that time.

The villagers are able to resize the projection of shadows of a tree in the planting period, for this they need to know the thickness and length of the branches, the size of the leaves this is the first thing they should know. Then you must see where the tree is going to bend if it is for *utchene*[90] or *mumphelagambo*, there yes, it is already possible to know if it will have shade in the ideal place to be used inside the yard of the house.

Shadows also play a crucial role in fisheries conservation. In summer, when there are droughts, one of the alternatives for survival is fish, made on the coast or in the creeks near the village, preferably under a shade over or in places with vegetation, because it is in very cold water that the fish appear, there is abundance of plankton, flowing to eat the leaves and branches that rot from the tree.

Shadows are also useful for fish conservation, i.e. when the fish is taken out of its habitat it urgently needs to be put in a sunless place, this is when fishing takes place during the intense sunshine, but at night one must be careful not to expose it to the moonlight so that the fish does

---

[90]In Xitswa language it means solar spring.

not get crushed, this is when the shade of the moonlight or the sun becomes useful to conserve the fish, especially when it has not yet passed its preparation, i.e. the intestines, scales and other parts that accelerate the process of putrefaction or rotting. Still on this subject, one of the residents added,

> [...] there are times of the year when trees begin to lose leaves, especially in the months of changing the foliage of the plants, this time it falls with the hibernation of some animals, the snakes even change their clothes too. This is a very good time to fish on the river banks under a tree because that is where a lot of fish stay who want to take advantage of everything that falls and rots as food [...]. (Verbal information)[91].

In this perspective, shadows and trees are associated with the existence of a number of smaller organisms which make up the plankton, essential for the aquatic animal food chain, which intensifies the formation of shoals of fish, making these places ideal for fishing. Knowledge of shadows is often ignored or goes unnoticed in people's minds, and there is a need for resignification as a way of drawing attention to their value in everyday life.

---

[91] Subject H.F.C. 56 years old, village of Aqui in Massinga, Inhambane, Mozambique.

CHAPTER II

57

PRODUCTION OF LIFE THROUGH SHADOWS

> *Knowledge of reality is light that always projects some shadows. It is never immediate and full. The revelations of the real are recurrent. The real is never what one might think, but it is always what one should have thought.*
>
> *Bachelard*

This chapter dealt with shadows as cultural, economic and political matters of daily life. In the city of Macapá, AP (BR) and in the village of Haqui (MZ), the equinoxes and solstices symbolize the overlapping of shadows, as well as the changing seasons that subsidize festivities for these events. Shadows are often understood as dark places that relate to the dark, understood as places of fear, strangeness, frightening and little is revealed of their potential in everyday life. An attempt has been made to bring the category of shadow and its contributions into different spheres of socio-cultural, economic and political life. The case study was based on the inhabitants of Aqui village in the district of Massinga in Mozambique (MZ) and the city of Macapá in the state of Amapá in Brazil (BR).

Shadows in life

The word shadow aggregates several meanings, and symbolizes knowledge that forms the basis for the formulation of knowledge, especially when understood as images projected on what is real. In this idea, they constitute abstractions that allow the recognition of objects and things through concepts, theorizations, and the production of synapses that make possible even absent objects and things, their identification through codes of language, making the shadows the images that approach a given reality about things.

Knowledge about shadow behaviour is necessary because it allows to understand social representations, for example it allows to understand the definition of sex in eggs of reptiles and turtles. It also influences public health planning, agriculture, livestock, urban planning, calendaring of festivities, transforming shadows into environmental variables of various utilities, being animal shelters and ideal places for "[...] hunting, looking for fruits or nuts and roots [...]" (MILONE, 2003, p. 9). It also helps in the preparation of calendars for various uses, such as marking seasons for planting, cleaning weeds, harvests and festive dates. Festivals generally coincide with the position of the shadow during the "[...] solstices or equinoxes [...]" (ALVES, 2006) corresponding to the months of March, June, September and December. In this case, shadows are not only an environmental variable, they are part of the life of the people of the place, representing the lived world, symbolically represented by the social structures organized by the use and use of shadows in daily life.

In the city, where temperatures are high, especially in summer, shadows play an important role in reducing "[...] the emission of pollutants released by generators from the energy used in environmental conditioning equipment in homes and offices [...]" (BELCHIOR, 2014, p.8). Together with evapotranspiration, shadows contribute "[...] to the cooling of the surrounding area by attenuating solar heating under artificial surface coverings of buildings, and these effects together manage to reduce the air temperature by more than 5°C [...]" (AKBARI *et al*, 1990 apud BELCHIOR, 2014, p.7). Although trees cause the air temperature to cool down in summer,

> [...] this effect will certainly be influenced by the type of ground cover, the spacing and arrangement of the trees, and the type of tree species-dimensions (height, crown and shape of the tree), the period of foliation (time of the year when the trees present themselves with or without leaves) and their shading coefficient [...]. (AKBARI *et al*, 1990 apud BELCHIOR, 2014, p. 7).

According to Snow (1982), thermal comfort is the mental state that expresses man's satisfaction with the thermal environment, which surrounds him playing a role of thermal neutrality[92]. The main factors on which thermal comfort studies are based are: the search for human satisfaction in relation to thermal variations; the need to improve performance in activities when done at pleasant temperatures; the desire to save or conserve energy by reducing waste with heating[93] the search for instinctive and cultural mechanisms for heat protection by avoiding exposure to the sun through natural and artificial shading among others that allow temperature reduction through ventilation, refrigeration, creation of microclimates or by regulating air temperature, relative humidity and breeze circulation.

Many African schools use the shadows as places to study, making it important to know their trajectories throughout the year, in order to know the ideal place to put, for example, a "shadow room"[94] of classes at a certain time of day and time of year (PHOTO 20).

---

[92] Physical state, in which all the heat generated by the organism through the metabolism, is exchanged in equal proportion with the surrounding environment, there being neither heat accumulation nor excessive loss of the same, maintaining a constant body temperature reinforced by the external environments since the organism does not have enough condition for a person to be in thermal comfort (SNOW, 1982).
[93] Vaporisation or evapotranspiration which occurs due to intense environmental heating, causing dehydration of the cells (MARTINS; GONZALEZ, 1995).
[94] Denomination of tree shade used as classrooms to study.

Source: Dadivo José (2016).

In the scenario illustrated in Photo 20, the shade is projected by a tree with a spherical shape, and its projection allows all students to embrace the shade with good shade in the months of January to July, from August to October there is leaf abscission leaving the shade room as a warm place and difficult to teach.

Based on Tuan (1980) it can be said that shadow is a culturally representative phenomenon of geographic space, that from it the inhabitants of the village of Aqui and the city of Macapá produce configurative knowledge of the lived, mental and conceived space, that is to say, this phenomenon symbolically represents cultural, economic and political issues of daily life.

Culturally, the inhabitants of the Povoado de Aqui in Mozambique consider the positions of the daytime shadow corresponding to the phases of human life, and when a person is born the shadow that is always present in that individual's life appears, when he dies, he disappears. They believe that the afterlife is established according to the type of shadow of the deceased, because if the person was bad during the earthly life, his shadow will turn into a ghost, which will torment the lives of the inhabitants in the dead of night. This realization is a teaching to practice good while alive in order to prevent his shadow from suffering in the postmortem. In this perspective, the shadow configures itself as a soul of the individual, but it also constitutes an element of moral and environmental education.

Culturally, it is possible to affirm that in Macapá the shadows, particularly of the equinoxes, represent dynamic moments of social structures in that they represent moments of changing seasons, symbolised by the festivities at the landmark monument zero. The equinoxes represent the zero shadows, which is perhaps why the Secretary of Tourism has been trying to organise various activities at the Marco Zero Monument, in an attempt to make residents aware of their importance. Attentive residents of the city can see that it rains a lot at the equinoxes, there is flowering, leaves fall from trees and the shadows are very short during all the months of the equinoxes.

To understand this relationship means to think of the shadow as a phenomenon that visualizes important moments to structure day-to-day activities. This mentality of shadows is subjective, its judgment depends on factors of order (biological, cultural, temporal, currents of thought and environment) involving the person, which from the brain perceives at the cognitive level and his imaginary. But the group plays an important role in the construction of the meaning of these shadows as positive or negative spaces. Thus, shadows can symbolize places that define animal sex, sacred, political power, leisure, problem solving, initiation rites, conflict and peace planning, etc.

To confirm this perspective a fellow from the village of Aqui describes that:

> [...] the shadow gives birth to many males of turtles, crocodiles and snakes; the shadow when it points to the position of the sunset represents death, so we bury the dead facing west; some shadows chill the whole body just by sitting there. You know that this place is of the dead or may even be of the wizards, only they are invisible but the shadow gives a signal; you may even feel warmth in a place with shadow [...]. (Verbal information)[95].

Based on Head (1995), Carvalho *et al* (2002) and Dias-Filho (2006), it can be said that the shadow is also a geographical physical phenomenon that can be conceived or planned intentionally as a target of intellectual imagination or not. The shading of the coast and riverbanks serves to spawn reptiles, turtles and birds, decreasing the average environmental temperature, causing an increase in the hatching of male offspring, which generates balances or imbalances between the male and female populations.

The shadows can be visualized by means of civil engineering projects designed to alleviate the high temperatures in urban and rural spaces, ventilation for animals and plants, among other activities related to the compensation of daytime thermal amplitudes.

The lived shadows, correspond to the place inhabited by humans including their plants and animals, the place where in the day to day they interact, with the variables of the environmental system of the place. It is equivalent to the real system of environmental representations, constituted by natural and artificial shadows. Which mark the space lived from the representations of autopoiesis and resilience, with boundaries marked by the feeling of possession of the inhabited space. It is the shadow of the backyard or residential area, as well as the others of the place frequented and well known by the residents.

---

[95] Subject L.A.B., 58 years old, village of Aqui in Massinga, Inhambane, Mozambique.

The inhabitants of Aqui emphasize that this lived shadow, serves both the residents of the house and the animals of the family and the jungle, as quoted below:

> [...] at home there must be shadows for all this, it is good to have a place for the animals, another for the children to play, another for the man of the house to receive friends to talk to; but there must always be a tree for *Kumpahla;*[96]there are shadows for snakes, rabbits, gazelles, etc. in the community. [...]. (Verbal information)[97].

In a broad perspective, the shadows symbolise the space of representations, marked by the offices of work, leisure and eternal rest spaces. They also represent the sundial that visualises the maximum declination of the sun to the north and south, occurring at a certain period of the year, being known as equinoxes that mark the passage of the sun on the plane of the Earth equator (SILVA; CATELLI; GIOVANNI, 2010, p.25).

The control of equinoxes and solstices by means of shadow has always been the subject of human observation, and for this reason Milone (2003, p. 9) states that the peoples of antiquity (Chaldeans, Phoenicians, Incas, Aztecs and Egyptians), held feasts to commemorate, the "[...] date on which the day beat the darkness [...]", December 21/22, corresponding to the summer solstice. The date of the winter solstice, with the emergence of Christianity, was decreed by Pope Julius I in 350 A.D. on the June feast, the summer solstice, reserved for the birth of the baby Jesus Christ, which was celebrated on December 25.

2.1 Shadowing and knowledge building

Shadows have never enjoyed a good reputation due to the lack of light, being associated with uncertain, strange things, wrapped in suspicions, understood as bad company, and therefore associated with what frightens. That which is inferior and dangerous, that which deceives, saddens and threatens. They have always been understood as disquieting phenomena, weird ones that: they grow and diminish, appear and disappear that are attached to the body, but there is no way to capture them (CASATI, 2001, p.117).

The same author states that although perceived as negative, they have played a crucial role in building scientific knowledge. It was at the basis of the shadows that Eratosthenes (276-194 BC) determined the size of the Earth by deducing that its shape was spherical when observing the shadow of two objects situated in distant places (COSTA, 1999). In ancient Aswan Siene and Alexandria in Egypt, crossed by the imaginary line of the Tropic of Cancer, during the June or winter solstice, at local solar noon, they found that the shadows overlapped the objects that projected them.

---

[96] In *Xitswa* language, it means ritual of evocation of ancestors.
[97] Subject L.A.B., 58 years old, village of Aqui in Massinga, Inhambane, Mozambique.

Eratosthenes' thought not only guided knowledge that allowed the determination of the shape and size of the earth, but also studies of eclipses, establishment of distances between earth, moon and sun, discovery of the structure of the solar system and the orbits of the planets. It allowed Galileo to deduct the relief of other planets and the moon, invent the linear perspective and understand the mathematical projections.

The shadow becomes "[...] a friend of knowledge by joining a club of shadow members [...]" (CASATI, 2001, p.17)), which brings together thinkers of antiquity, the Middle Ages and the Renaissance, such as Galileo Galilei (1564-1642) and Ptolemy (90-168), whose rediscovery is reborn with the paintings of Leonardo da Vinci (1452-1519), driving forward two great advances in the Western world, creating the geometric perspective that allows objects to be inserted into space during drawing or painting.

The shadow has become important for reading the background and perspectives of an image, giving relief to images such as making up the face that needs to increase the shadow to give more depth to the face. In the modern age, scientific studies related to the theory of shadows, sombrography[98] and improvement of a sundial useful for planning daily activities have become popular, all based on knowledge about shadows.

In urban planning, they have come to play an important role, as they are essential in the orientation of houses and streets of the city, avoiding obstruction of the sky or making the city excessively dark by the deficient sunlight. In New York City, Manhattan's skyscrapers with giant shadows make it difficult to sunbathe all over the block. Making urban planning crucial by means of readings and interpretation of the trajectory that the shadows perform throughout the day in a given geographical area, that is, the builders of a city must have the domain of the "[...] anti-shadow regulation [...]" (CASATI, 2001, p. 28).

During the Middle Ages, studies on the shadows remind us of the writings of Arabs such as Al-Burini who wrote a book called 'Complete Treaty of Shadows' between 1030-1040. Determining the time when the shadow is equal to twice and half of things as times to recite Islamic prayers, in this case the shadow of things and objects constitutes a sundial to determine the time when Muslims should recite their prayers throughout the day.

Philosophers and Renaissance scholars used the concept of shadow to analogically mean two movements of thought: wisdom and ignorance. In the book the Republic, Plato (427-347a.C.) (2001) drives the capacity for abstraction, starting from the allegory of the cave, comparing the condition of ignorance as a shadow or darkness and that of reasoning to attain wisdom as the moment of illumination.

This phase is considered by the Enlightenment[99] and brought together thinkers such as

---

[98] Direct technique of exposing objects on photosensitive paper.
[99] An intellectual movement originating in France defended the human capacity to know the world through science, considered an expression of rigor, objectivity and predictability. It exalted that, through science,

René Descartes (1596-1650), precursor of methodical doubt, John Locke (1999b) of empiricism, Voltaire (1694-1778) defender of freedom of thought, Rousseou (1712-1778) who wrote about democracy and equality, Montesquieu (1689-1755) who instituted through his book the need for the division of powers in any State. The concern of these thinkers was to propel minds from the world of shadows into the world of lights. By means of theories that enlightened the Europeans, particularly the French, shaken by the contradictions of the old regime under the yoke of a lengthy, sharpened land system, which generated dissatisfaction in the various sectors of society, especially among the bourgeoisie and the small peasants.

To leave the world of shadows for the world of lights is an expression that symbolizes the conceived and mental shadow, used to mobilize reasoning that would allow popular revolts, to highlight the French revolution that propagated the exit of the individual from shadows or ignorance to the world of reason and the search for new revolutionary knowledge. Thus, the light represents what would drive the French revolution and all the European revolutions of the time to break out. It subscribes to itself as knowledge, philosophical and scientific, which has turned thought into clarity, in the revealing and enlightened sense of the human personality in the search for knowledge and knowledge that is constituted as the antonym of ignorance or shadows.

According to Baxandall (1997) and Casati (2001), Cristovão Colombo's (1451-1506) circumnavigation trips were facilitated by means of geographical orientation based on the domain of the trajectory that the shadows of objects and things project, indicating the direction and direction serving as cardinal points.

Nowadays cognitive psychology, astrology and the philosophy of science rely on shadow analysis for their object of study. They become precious instruments of knowledge, which "[...] instead of hiding reveal the [...] light of what to do wrong [...]" (CASATI, 2001, p.9). Making the history of science the product of shadows, this is because science does not represent concrete objects, but rather the abstraction which brings us closer to reality. Psychologically, the shadow is the interior of the individual characterized by undesirable experiences "[...], therefore, the only way to know this side and help him to heal himself is to know the shadow [...]" (SALETE, 2009, p.138).

For the psychoanalyst Jung (1875-1961) (2011) the shadow is the archetype of the ego of the individual and darker, that is, the shadow constitutes the animal-like part of the human personality, inherited from the lower forms of life from when we were savages until reaching the present stage. In this order of thought, the shadow contains all those activities and desires that can be considered immoral and violent, those that society and the individual himself do not approve of

---

man could scare away the fear caused by *shadows*, ignorance and superstition, guarding the hope of a world where the lights of reason would allow the best possible quality of life and the emancipation from prejudice, violence and arbitrariness. According to the enlightening philosophers, this way of thinking had the purpose of throwing *lights into the darkness in which a* large part of humanity found itself.

or assume as normal for his conduct as a human. And when this occurs, it is claimed that the individual has been affected by something that was beyond his control. This something is the shadow, the primitive and savage part of human nature which "[...] seems to ask for some kind of intervention, perhaps a therapy, or pill, perhaps a trip to the confessional, or a confrontation with the soul at midnight. As soon as people recognise having it, they want to get rid of it [...]" (CHOPRA *et al*, 2010, p.9). Our shadows represent what no one wants to be, but unfortunately all humans have hidden feelings, those that constitute misbehaviour, such as the desire to surrender to addiction, to explode, to quarrel. When discovered, they allow the human being to purify himself to become kind. This is what makes the shadows into psychoanalysis primordial representatives of physical things, visualized in reality as "[...] images of the soul [...]" (CASATI, 2001, p.46)

According to Tuan (1980) and Chopra *et al*, (2010), shadows can also represent dark aspects such as fear, anger, anxiety and violence, among other demonic possessions, whose solution is to purify the affected person through rituals, cleansing of the body, fasting and exhaustive austerities. Thus, the shadows represent the interior of the individual and all negative behaviour, and it is necessary to subject this person to psychological therapy. In the Brazilian Amazon, there is a spiritual practice which seeks to read and interpret the dark part of the individual through *ayahuasca*. In this perspective, Medeiros (2016) states that the use of Ayahuasca within an appropriate ritual, functioning as a therapeutic agent, opens up the possibility of accessing internal contents that are treated as shadows, generating physical, mental, and emotional progress of the drinker.

Still based on Medeiros (2016), one can say that Ayahuasca is a hallucinogenic drink prepared with the stem of *caapi* or *Banisteriopsiscaapi* and leaves of *chacrona* or *Psychotriaviridis*, ritually produced by sects, such as: *auasca, daime, ygê, mariri, uasca,* professed by Amazonian populations, with followers in other parts of Brazil and the world. In other words, it is a spiritual drink that brings from the unconscious all that is abstract, that is, the shadows, which are difficult to read because they represent themselves colorless, boring, perplexing and in expanded or diminished form of objects that project them, becoming absences and negative things.

The shadows for being represented in an enlarged or diminished form of objects induce the movement of thought to approach reality, making them knowledge-friendly, for it is this that brings us closer to reality by means of scientific methods and techniques. That is to say, in scientific production one never sees the real, but rather reaches its shadow that constitutes reality and not the real itself.

According to Almeida (2006) and Axt (2007), in physics shadow represents phenomena of geometric optics, evidencing the rectilinear propagation of light and air, resulting in the formation of shadow, which occurs when light finds an opaque object in its path (every object that does not allow the propagation of light through itself). In this case, the shadow results from a "[...] local and relative deficiency of the amount of visible light [...]" (BAXANDALL, 1997, p.17), with light depending on the variation of the shadow to illuminate, becoming an essential element for light to exist.

Light is the quantity of the flow of the mass-energy unit emitted by a source of radiation, the sun or a candle, but it is the shadow that gives volume and authenticity allowing visualization of images, so they are two physical phenomena that complement each other, so neither exists without the other. Therefore, shadow is a phenomenon that gives existence to light and light that gives rise to shadow.

Baxandall (1997) quoting Leonardo da Vinci (1452-1519) states that there are three types of shadows: 1. the shadow connected or self shade is that which is on the object which produces it; 2. the shadow cast which is detached on another surface and is characterized by being intense on the projected surface; 3. the shadow or oblique and sloping shadow is the half shadow which is therefore weak or of low intensity on the projected surface.

The connected shade does not obstruct photons[100] of light on other objects, it is suitable to be used in thermal comfort because it does not project out of reach and allows rationalization of spaces in case of small surfaces while the cast or projected shade requires more geographic space and can serve for more people who take advantage, despite this advantage, is not always sustainable because it projects out of reach or rational use.

All objects (solids) project shadows, automatically represented by their shadow, placing the shadow in an environmental dimension, referred to by the theoretical paradigm of systemic complexity, which sees the environmental system as a relational set of system variables, planet earth consisting of the atmosphere (air), lithosphere (soil), hydrosphere (water) and the biosphere (living beings) including man himself and other inanimate beings (ROHDE, 1996). In this perspective, the shadow becomes a complex and systemic socio-environmental variable that can be explained in several ways: physical phenomenon, resulting from illumination or light and responsible for vital cycles; an analogical metaphor of knowledge and everyday knowledge; an element that helps the reconstruction of the three-dimensional form of objects that projects them by calculating the height of larger objects, allowing the perception of the geometric beauty of artistic images among others. In other words, the world without shadows would not be funny, because they would not have the thickness and delimitation of the background of things, since it is shadows that define the complexity of the environmental system in the model we see. Thus, the shadow abandons the place of opposition to light, becoming the complement of the solar radiation and the opacity of the objects, becoming at the same time a dynamic element of life.

---

[100] These are surpluses of smaller particles of energy which come together and become larger particles in the form of a particle disturbance or wavelength illuminating or making light (BAXANDALL, 1997, p. 17).

Its driving role has always been recognized by all peoples more than 2000 B.C. by guiding the Egyptians to build the pyramids (Kéops, Quefren and Mikerinos) in "[...] north-south and east-west directions, with an average error of about 1/22, 5 degrees for each direction, as guidelines for the geographic poles of the earth to be used to measure shadows and determine the arrival of equinoxes [...]" (RODRIGUES JÚNIOR, 2012, p.56). Also at Stonehenge in the United Kingdom, solar shading was applied to demarcate "[...] points of solstice, equinoxes and the maximum and minimum declinations, although due to the precession movement some alignments no longer mark". (RODRIGUES JÚNIOR, 2012, p. 49).

Knowledge about shade has various applications in everyday life, such as identifying colours of clothes that absorb less heat, regulate agricultural production through shadow farming and conservation of cultivars, and serve in fishing, as is the case with Ba tswa people who identify shoals of fish and fishing periods due to shading, among other elements.

Such statements were also captured in the interviews, as highlighted below:

> [...] the city is known to be very hot from August to September, so no one is advised to wear black clothes because they absorb heat; at that time even black clothes are on sale because no one needs them [...]. (Verbal information)[101].

> [...] we all know the behaviour of animals. For example: fish, rats, partridges, reproduce a lot in *nhahure, ndzati, nlanguloehukhuri* time, (August, September, October and November respectively), when the shadows overlap in December we already have in abundance [...]. (Verbal information)[102].

It is the shadows that allow the prediction of light (CASATI, 2001), without it, one can hardly understand the opaque and dark bodies, the interior and its limits. Night is the largest shadow projected by the earth faithfully representing a dark place full of a complexity of environmental variables. Hence the need to consider them, the most complex variable that allows reading, interpretation of the light and the background of all the environmental complexities represented by the earth system.

According to Plato (427-347 B.C.), everything that grows, reproduces and feeds has a soul that needs to be respected because it constitutes the psychic or active principle of life. Therefore, Aristotle (1995), plants, animals and human beings would have soul visualized through the shadows projected by these living beings. For Thomas (1979), shadows are proof of the divine manifestation and are mentally immortal, serving also in the connection of men with God by guarding the essence of human existence. Shadows are projected representations of plants, animals, human beings, of all things and objects, therefore, the soul of these things and objects that project them.

By attributing the role of soul it ascends to the category of immortality, and physically

---

[101] Subject J. A. M., 45 years old, ordinary citizen, city of Macapá, Amapá, Brazil.
[102] Subject A.C. D., 50 years old, village of Aqui in Massinga, Inhambane, Mozambique.

and visibly represents "[...] the soul of things [...]" (CASATI, 2001, p.36), playing a vital dynamic. On the basis of Gilson (2006), it can be said that etymologically the term soul comes from the word (anima or animus), which in Portuguese generally refers to the animating principle of all living beings, be they animal or vegetable.

For Plato (2001) and St. Augustine of Hippo (AGOSTINHO..., 1987), shadows are immaterial, just as the soul which is man's immaterial, life does not depend on the body, it depends on the soul which is the beginning of life. Men are not the only beings who possess soul or psyche; all living beings, including bodies in general, possess the soul that is represented in real life as the shadow that is born of the individual, or the thing.

The alternation between light or light and darkness, night, constitutes a primordial element for the construction of a vital concept called day equivalent to twenty-four (24) hours or to "[...] the first notion of time [...]" (MILONE, 2003, p.16).) in this case, darkness is the shadow which symbolises the vital cycles, the period of night and sleep, hibernation, rest, the recharging of energy lost during sunlight while sunlight at waking time, the solar irradiation necessary for photosynthesis, the distribution of mineral salts and energy for plants and animals.

When understood as the soul of things and objects, shadows are attributed beyond movement, but also a transcendental place at the top of the vital hierarchy, such as immortality or continuity of life after death. Allowing constructions of convictions, like that of the Chinese who think that if the shadow of an untouchable one touches the body of a Brahmin, it will fill it with impurities, and there will be a need to purify it. Cautioning the gravediggers not to let the shadow slide on a coffin or a grave, by tying a ribbon over his shadow. Australians do not allow the shadow of one person in movement to come into contact while the other person is asleep because the person may wake up seriously ill.

The Ba tswa, believe that if the enemy stings the shadow of his rival with a sharp object, then he will sicken and die. Just as the Yoruba of sub-Saharan Africa believe that one can harm someone by doing work or magic on their shadow.

For many societies, the longest or longest shadow corresponds to the bravery of living beings, the good mood, the moment when the individual considers himself stronger. This moment is registered during solar chance or in the first hours and at the end of the day, corresponding to the period of mild temperatures. It is the moment of thermal comfort, of good disposition to work, mainly in the primary and secondary sector (agriculture, fishing, mining, etc). The short shade, which happens around midday local until about 15 hours and 30 minutes, corresponds to the hottest period of the day, characterizing a weak man due to the decrease in strength, creating weakness, laziness and providing a pleasant environment for rest.

For family farmers Ba tswa and Zulus[103] the short shade represents the moment of greatest sunshine, very high temperatures and thermal discomfort, symbolizing the death of plants that wither, dry out while the animals become unwell.

Shadows are the most important vital part of bodies, if they constitute like the heart, and there is a need to study and preserve them, because they constitute the visible form of the soul of beings, things, objects, and through them recognize their traits. In this sense, reflections or perceptions about the importance of shadows in geographic space are relevant, since they constitute the visible form of the soul of beings, objects, and through them can characterize traces in a given context.

According to Morgan (1996), Hertz (2005) and Santos (2009), shadows are the reflection of the existence of light. It is important to recognize traces as a reflection projected through this thing or object "[...] present in all its moments of life, and which only disappears definitively when the organism dies, constituting a form of existence [...]", constituting immortality and life after death.

Theologically, the shadow represents the Lord Almighty, the creator of all things visible and invisible (God), in Genesis (1:2) when the earth was formed, it is reported that the earth was empty and vacant with darkness covering the abyss, that abyss is the shadow. That magical powers are attributed through healing to the crippled by touching the miraculous shadow of St. Peter. In the Bible (1980) one quotes the sun clock of Ahaz. In 2 Kings (20:8-11) and Isaiah (38:8), when Hezekiah said to Isaiah: "What is the sign that the Lord will heal you, and that on the third day you will go up to the house of the Lord? Isaiah answered that: The shadow will go forward ten degrees, or go back ten degrees. In this context, the shadow symbolizes a living being, especially a very powerful human being who heals diseases and helps in the vital dynamics of all living and non-living beings. At the same time, the shadow represents a geographic space built by the "omnipotent, corresponding to the religious environmental system, with the presence of the Lord, the shadow representing hell, darkness, the dark, dangerous place while light to paradise and eternal happiness". (GENESIS, 1:4).

Still on the conceived representation of the shadows, the choreographer LinHwai-Min and the Chinese artist Cai Guo-Qiang, created a contemporary dance work called Shadow Dance or "Wind Shadow", where the shadows come to life, moving through the use of black and white monochromatic palettes and the contrast between light and shadow, giving life to the dolls with sound and light effects, creating the impression that they are humans telling their life stories.

Because shadows symbolize the soul of things, and all beings have souls that continue their life after death, the living need to care for the dead. Thus, in the city of Macapá and in the

---

[103] They are peoples living in a part of Mozambique, South Africa, Lesotho, Swaziland and Zimbabwe. Today, they have restricted political expansion and power, but in the past they formed a warrior ethnic group that resisted the British and Boer Imperialist invasion in the 19th century.

village of Aqui, the inhabitants make shadows in cemeteries, in the belief that the grave needs to be kept in a comfortable environment because it preserves the physical body of the deceased and their soul needs tranquillity and peace. This belief adds to the cemetery the category of the last dwelling or eternal home of human beings, marking psychological time very different from the absolute and true time of Isaac Newton (1979) measurable by means of the mechanical clock.

Starting with Plato (427 - 347 BC) and Aristotle (384 - 322 BC) it can be said that the shadow is both a sundial, a place of thermal comfort which preserves the grave of the deceased from thermal variations and because it is the soul, it is also the driving machine for reasoning. As a sundial, it moves or has daytime mobility, allowing men to number or schedule, perform various activities and social convictions while, as a reason, the shadow/soul is responsible for the construction of knowledge.

2.2 The socio-environmental value of cemeteries through shadows

For VovelleViollet-Le-Duc (1993), cemeteries are privileged resting places, rustic places full of monuments capable of receiving all the tributes of family memory and civic respect, making the tombs "[...] vast fields for the study of archaeology, ethnology, history, arts and philosophy. In this perspective, the cemetery is the second residence of the deceased, where the tomb is the house, full of architectural and landscape elements present in everyday life or in the world of the living"[...] it is there where the socio-economic order of the living is reproduced in fact or in an idealized way". (RAGON, 1981, p.37). Transforming cemeteries into socio-environmental heritages embodied in symbols and values which will give meaning to life.

This socio-environmental value is related to the fact that cemeteries represent areas of shading through awnings or roofs for tombs and arboreal or green, because they are places of permanent abode of the dead and used by the living to commune with the spirits or souls of their ancestors. As one resident of the city of Macapá stressed:

> ...] the church has taught me that the dead are not dead, they are alive, with another life, this is written in the book Wisdom (2:23) of the Holy Bible; that is why we have the obligation to put our dead in a comfortable place with a good shadow, to cleanse, to care for [...]. (Verbal information)[104].

Analysing the placement of Wisdom (2:23) it is possible to affirm that the idea of caring for the tombs of the deceased seeks its inspiration in the catechism of the Catholic Church, which teaches believers that all those who die in the grace and friendship of God are not always completely purified, although they have been guaranteed their eternal salvation, and will therefore, after their death, undergo a purification in order to obtain the holiness necessary to enter into the joy of heaven, thus shading and giving other care to the tombs creates joy and a

---

[104] Subject G.M., 48 years old, ordinary citizen, city of Macapá, Amapá, Brazil.

pleasant environment for the deceased.

> [...] people try to get a shadow to bury their deceased, but it is difficult to get ground for burial under these hoses, only rich people who have made a reservation; it is comfortable to have a relative buried there, because the tomb will not get sun, besides, the relatives will have shade when they come to visit [...]. (Verbal information)[105].

Besides the shadows being comfortable for the deceased, there is belief that the individual when dying sanctifies himself, being able to control or perform miracles in the lives of his relatives who are still alive creating comfortable environments in the tombs of his deceased relatives in order to obtain some grace.

> [...] it does not have to be only candomblé to believe in the active existence of one's ancestors and accept that death represents a change of life for another; the dead free themselves from all the restrictions imposed by the land, thus acquiring potential that can be used to benefit their relatives who are still on the land. For this reason, it is necessary to keep them in a state of peace and contentment, as happened with ÒkúÒrun and *Àgbagba* who continue to communicate with the living until today, protecting and caring for our destinies on earth, so it is important to care in a pleasant way for the tomb of the relatives in places with a good environment like these canvases and hoses [...]. (Verbal information)[106].

The shading of cemeteries is one of the priority items in social and environmental planning, because these places constitute the cultural heritage and very frequented spaces, constituting consolidated tourist points in the most different countries of the world according to Viollet-le-Duc Vovelle (1993). Supported in Charlet (2003), one can affirm that by attracting visitors from all over the world interested in knowing the tombs of personalities from different areas of knowledge, appreciating works of art that decorate the tombs or simply enjoying moments of peace and tranquility in the wooded gardens turning these spaces into a cemetery-garden, open to the public. As L.B.A. also points out (resident of Macapá):

> [...] planning a city is very complex, the mayors need to have councillors and a complex and well prepared team to understand cemeteries as places that preserve stories of people who have had impacts, that inspire positively or negatively, they must understand that they are places of curiosity, places that attract visitors, so it is a priority that in a city like our Macapá that is very hot, seek to enhance shadows, allowing visitors to do their research in a humanly comfortable environment, I always thought so; It happens that our cemeteries have no structure thinking in this sense, it is sad that when we go there we suffocate from so much sun, we burn the whole body even applying sunscreen, this causes skin diseases in the residents of the city [...]. (Verbal information)[107].

---

[105] Subject F.L.S., 41 years old, ordinary citizen, city of Macapá, Amapá, Brazil.
[106] Subject A.B.C., 35 years old, ordinary citizen, city of Macapá, Amapá, Brazil.
[107] Subject L.B.A., 47 years old, professor at UNIFAP, city of Macapá, Amapá, Brazil.

Osman and Ribeiro (2007) cite some examples of good cemetery structures with shading, creating a comfortable environment for visitors in Brazil, as well as, on the tourist route around the world, starting with the French, PèreLachaise, Montparnasse and Montmartre, followed by the English cemeteries: Highgate and Golders Green Crematorium in London. In South America he leads the Recoleta cemetery in Buenos Aires, followed by the Brazilian cemeteries, Consolação and Morumbi in São Paulo, São João Baptista in Rio de Janeiro.

Another Brazilian example is the comfortable environment produced from tree shading in the Cemitério do Imigrante, constituting "[...] one of the few Brazilian cemeteries erected in the landscaping of a forest, conserving and cultivating leafy trees by selecting regional decorative plants [...]" (VALLADARES, 1972, p. 310). These places are environments of tranquillity, rest and cultural rescue provided by shading and green spaces that represent an encounter with what is natural, with ancestry, symbolizing spring, youth and passion, fertility, development, wealth, good luck, hope, fresher environments.

According to Trees[2016) as it is a priority today to transform cemeteries into fresh and pleasant surroundings, projects are emerging to plan sustainable cemeteries, highlighting the Capsula Mundi project in Italy, which consists of the burial of the deceased in an organic and biodegradable egg-shaped capsule containing the roots of a seedling or seeds which is capable of transforming a decaying body into nutrients for a tree which will transform cemeteries into forests or sacred shading spaces.

Similar experiences are being developed in Brazil, such as the Horto da Paz woods or cemetery in São Paulo (FIGURE 58) with a small wood that received the name Paz e Vida (Peace and Life) has at least 300 trees that the relatives plant at the moment they leave the ashes of the deceased next to the seedlings. The relatives who choose to cremate acquire an ecological capsule, made of totally biodegradable coconut fibres, which is accompanied by seeds or a tree seedling to be planted together with the ashes to form a forest that projects sacred shadows. In addition to the arboreal shading, there are protective heat screens on the graves such as the cemetery Nossa Senhora de Conceição de Macapá.

The top cover of tombs is done in an East-West orientation, allowing the shade to be superimposed throughout the year at the hottest time of the day. Usually it is the relatives of the deceased who have the sunscreen built, doing frequent cleaning taking advantage of the shade during visits to the place.

Some testimonies from the residents of Macapá about the shading of cemeteries:

> [...] it is inconceivable that in a very hot city like Macapa, there is no ventilation system to make the cemetery the least suffering place. The family members suffer for the loss of their loved one, they still suffer for being roasted

by the sun and when they return home they all have headaches, chills and possibilities of starting skin cancer, it is sad even [...]. (Verbal information)[108].

[...] like no one else in Brazil does for the people, the way is to turn around, here in the city we look for a specialist in putting on sun protection so as not to sunbathe Papa's tomb; so when it's the day of the deceased agent we even bring the ball and shirt of the Flemish team, we sit there right after praying for his soul; it's us who wins even with the shade, besides the fact that the person has put it on very carefully so as not to let the sun burn in any way [...]. (Verbal information)[109]. (PHOTO 21).

Photo 21 - Macapá tomb shading

Source: The author (2016).

The top cover of tombs is made following a technique that the shadow is superimposed throughout the year at the hottest time of the day. Usually it is the relatives of the deceased who have the sunscreen built, do frequent cleaning taking advantage of the shade during visits to the place.

2.3 Influence of shadows on human health

Solar radiation and shading play an important role in the physical and health maintenance of living beings and in particular of man. The sun's rays are important for life as they provide vitamins to the body, but can also cause disease when received in excess. From this perspective, the city of Macapá and the village of Haqui, which have thermal amplitudes, oscillating around 27°c, will receive throughout the year sunshine varying from 114 to 285 hours during the equinoxes and solstices. This will cause the two study sites to receive intense solar radiation, with

---

[108] Subject B.L.N., 35 years old, tourist in the city of Macapá, Amapá, Brazil.
[109] Subject F.F.M., 46 years old, city of Macapá, Amapá, Brazil.

a higher probability of exposure of their residents to sun-related diseases, as the next social subject states:

> [...] when it gets hot here in the village of Aqui, you can even forget that we have just come out of very cold weather; in the village of Aqui when it is hot you can heat water and bake mafura; we burn very well, I get very black in hot weather, worse now that many roads are running out of good shadows [...]. (Verbal information)[110].

> [...] at home no one goes out without an umbrella or a hat, everyone learns this from an early age; if fooling around here in Macapá is when the results appear, that's why you should learn from childhood how to protect yourself from the sun, every year in this city is a sun that does not end [...]. (Verbal information)[111].

Based on Conceição (2008), Fraser and Broom (1990), it can be said that humans, when subjected to a lack of thermal comfort, reach a stress situation, reducing their performance in daily activities, especially when the body receives intense sunlight at the hottest time of the day around 11:30 am until 3:30 pm local, causing thermal discomfort and health complications. Summer is a time of greater sunshine, a time when the inhabitants of the village of Aqui choose paths that have many shadows on their way to avoid getting too much sun. In this context the shadows guide the routes, that is to say the shade trees become route guides to the destinations. The felling of trees for farms or for housing construction has been one of the reasons that change the routes considered fast, making the destinations longer. Macapá is already a record holder for lack of shadows, almost the whole centre and other places that are opposites of the city, the pedestrian access is only good catching sun.

During sunrise and sunset, the temperatures are mild, at the same time as the projection of shadows is long or long, derived by the entry of inclined light, i.e. the angle of incidence of sunlight by means of oblique rays. In this period the fixation and metabolic process for vitamin synthesis called D in the human body is accelerated.

The best time for the body to synthesize Vitamin D is from 6 a.m. to 10 a.m. and from 3:30 p.m. to the sunset, because of the angle of incidence of the sun's rays. The organism does not synthesize solar Vitamin D after 10:30 to 15 hours. The time needed varies according to the time of day, the season and the location of the city. In this case, the further away from Ecuador, the greater its risk of Vitamin D deficiency.

Vitamin D, is acquired in the period when the projected shadows are long (in the morning), when the sun's rays when received by people provide the human body with a good physical appearance and good psychic health to the individual, contributing to the decrease in the incidence of depression among other diseases of the psychic apparatus, especially in the elderly,

---

[110] Subject E.F.G., 44 years old, village of Aqui in Massinga, Inhambane, Mozambique.
[111] Subject N.M.C., 23 years old, ordinary citizen, city of Macapá, Amapá, Brazil.

children and prisoners who often stay locked in homes without access to the sun at good times to enjoy the sun.

> [...] the sunbathing is only on the beach, but the doctors do not recommend for places like our city where at dawn there is already an oven, a burning heat [...]. (Verbal information)[112].

> [...] the good thing about our town is that even in hot weather, some mornings are cold, at sunrise it is good to catch the sun's rays to be in good mood for the day; already in cold weather we get angry with each other looking for a space to catch the sun's rays; in cold weather people's shadows can close the sun and create difficulties in catching the sun's rays; so as not to fight, we have the habit of staying next to each other to avoid blocking the rays through the projection of shadows [...]. (Verbal information)[113].

The morning and evening periods are good times for the body to synthesize vitamin D, therefore recommended for exposure on the beach. While the hottest period of the day, by coincidence the shadows projected are short, corresponding to the period not advised for exposure to the sun and can cause diseases, stress and other discomforts to the human being.

According to Pereira (2003) and Popimet *al* (2008), vitamin D is an essential substance for good cellular function regulating insomnia, bad mood and depression or chronic fatigue problems. Exposure at the hottest time of the day constitutes risks under ultraviolet rays (UVR) which are harmful to humans and are responsible for skin and psychic diseases.

> In this city I don't have much to talk about, but in Rio Grande Sul, the people there are depressed in cold weather, followed many suicides in the harsh winter below zero degrees [...]. (Verbal information)[114].

> [...] I don't like the cold weather because we are only confined at home always around the fire; now summer is something else, there are many places to go, there are many parties, it is a time of great joy, there are many tourists on the beach with money to buy and give jobs, it is very cheerful even in spite of the heat people move around more than in cold weather; in the heat it has more fruit also despite the fact that winter harvests are also very pleasant [...]. (Verbal information)[115].

Therefore, shadows constitute an essential socio-environmental variable for the protection of solar radiation, dangerous to the life of living beings, especially when they are used for thermal comfort exactly at those moments when the temperature performs thermal discomfort that coincides in the hours of short shadows, as mentioned above, they are short due to the angle of perpendicular incidence.

---

[112] Subject B.L.N. 35 years old, tourist in the city of Macapá, Amapá, Brazil.
[113] Subject X.V.M., 40 years old, village of Aqui in Massinga, Inhambane, Mozambique.
[114] Subject B.L.N 35 years old, tourist city of Macapá, Amapá, Brazil.
[115] Subject X.V.M., 40 years old, village of Aqui in Massinga, Inhambane, Mozambique.

Among several diseases influenced by lack of shading, Popimet *al* (2008), Miotet *al* (2009) and Hayashideet *al* (2010) highlight the occurrence of sunburn or erythema, mainly in children and fair-skinned people. They also refer to early aging characterised by the appearance of wrinkles and blemishes on the skin also known as photo-aging, especially on fair skin.

The same authors state that excessive exposure in places of higher sunlight can also cause vision problems due to corneal burns, causing cataract, pterygium and even skin cancer of the eyelids, herpes, acne (burning and redness), skin allergies, melasmas (brown spots that usually appear on the hands, arms and face), keratosis (rough and small wounds that hardly heal), skin cancer, among other dermatological disorders that have the ultraviolet rays as a triggering factor.

2.4   **The shadow as an environmental dimension**

Shadow as an environmental dimension represents the physical geographical phenomenon responsible for the obstruction of solar radiation, establishing microclimatic influences pleasant to the environment that allow the regulation of excess heat. Shadows of trees are the most recommended because they add environmental functions to the artificial ones, since in addition to establishing climatic relations between the inhabitants and the environment, they represent sources of oxygen, carbon dioxide filters, shelter for animals, and reservoirs of ecosystems, transforming these sites into excellent spaces for social coexistence and maintenance of public health, in addition to the ideals for carrying out diverse activities.

In the village of Aqui, shadows are elements of social division, separating the members of a family according to sex, age and function, that is, while women cook in one shade, children play in another, bringing them all together in the same shade when there is an obligation, which can be a sacred ritual, debate on a family problem (pregnancy of a daughter or relative, marriage/lobolo) or at the time of taking daily meals, when the mats are stretched out on the floor to sit the women and trunks or chairs for the boys and other male members (PHOTO 22 and 23).

Photo 22 - Shadows in everyday life - village of Aqui

Source: The author (2015).

Photo 23 - Shadows in everyday life - village of Haqui

Source: The author (2015).

The residents of Aqui commented the following about the shadows:

77

[...] we usually split up in the shadows at home, the children are in one playing with them there and we only get together when it's time to eat, there we all stay in the same shade [...]. (Verbal information)[116].

[...] every day when I wake up we all sit under a cashew tree that has at home to eat yesterday's leftovers like matabicho, then me, my children and my husband go to the machamba[117]; There I spread a capulana[118] in the shade of a mafureira[119] and leave my two youngest children to play while I and the other children cultivate; when the sun gets very hot around noon, we gather and join my youngest children in the shade for lunch, it can be with cassava or anything I have to eat [...]. (Verbal information)[120].

Eating together under the shade with other forms of use and exploitation of trees as places of social coexistence is one of the ways of recognizing the importance of the tree, especially its shade, for improving the environmental quality in the village. In recognition of this value, they select tree species as a form of species preservation.

Studies such as Sequeira (2014), Castro and Dias (2013), show that the city of Macapá, where temperatures are high mainly at the equinoxes, associated with the growing implantation of industry, commerce, rising population growth, causing disorderly expansion and verticalisation of the city, the use and exploitation of shadows tends to reduce, replaced by ventilation and refrigeration by means of air conditioners and fans. A trend that points to increased psychological stress, skin diseases, increased use of chemicals on the skin for sun protection, lack of love with plants for the young,[121]among others. The following is a testimonial from residents of Macapá:

[...] look my son, the three of us are construction engineers, but what we have seen in the cities of Brazil, including this one, is a lack of will of the people indicated to manage the city, which instead of tidying it up, are destroying it. For example, in Macapá the vegetation that should be used to reduce the solar gain throughout the day, continues to be planted in the way that each one understands, without obeying any criteria, the municipal body that should regulate this, does not realize the message; everyone puts whatever tree they want in the yard and on the pavement of their house, no selection of suitable species and planting on the east and west sides; there is no promotion by the city on the use of roof tiles and green walls for buildings, including the lack of habits of creating green spaces of undergrowth through gardens that can help cool the soil [...]. (Verbal information)[122].

[...] on the other hand, everyone goes crazy, looking for ways to cool down or buying umbrellas that are on sale at absurd prices, the rich always walk on the

---

[116] Subject H. I. J., 63 years old, village of Aqui in Massinga, Inhambane, Mozambique.
[117] Agricultural property, known as "roça" in some regions of Brazil.
[118] Cloth traditionally used by women to cover their bodies, carry babies, make skirts, trousers, etc.
[119] Mafura is the fruit of the mafureira plant typical of tropical Africa and abundant in Mozambique. Consumed after softening in water for 1h or longer, depending on the thermal state of the water or the atmospheric state of the environment of the place (C.P., 2016).
[120] Subject A. L. 38 years old, village of Aqui in Massinga, Inhambane, Mozambique.
[121] Lack of love with plants means that, due to the lack of incentives to live in wooded environments, new generations have begun to lose interest in planting, using and taking advantage of shadows from trees.
[122] Subject L.M.N., 59 years old, ordinary citizen, city of Macapá, Amapá, Brazil.

air conditioning so much at home, in the car and at work, while the poor burn in the sun, competing pavements with the cars or looking for the few shadows projected by the buildings, because in the city centre there are no shadows [...]. (Verbal information)[123].

[...] without shadows in the city, it has become a sad place, it no longer listens to that song of the birds that does good for the soul, when it is rainy season everything is quickly flooded, the green that brings joy to the view of the city, finally, we need to do a lot to make it pleasant for us and for visitors; the only place where there is vegetation that reminds us of the Amazon around here is only at the Federal University of Amapá, where the other places of our Amazon identity are [...]. (Verbal information)[124].

The narratives of the residents refer to the thought that the amount of free spaces of public use per inhabitant, corresponding to the Green Area Index (GAI) and the percentage of the Area of land occupied by Vegetation (GAP), is below that desired to consider Macapá a comfortably shaded city. From this perspective, writings by Belchior (2014), reinforce the need for tree shading, because it can decrease in the environment of a city, about five degrees centigrade (5°c) of temperature, especially when the trees are planted properly.

The afforestation must be adequate, following the standards required by the Brazilian Society of Urban Afforestation (SBAU) and the National Recreation Association of the United States, which recommend for a PCA of 18.40% and ICA of 31.70 m2 hab-1 b. Therefore, Macapá does not present in its entirety the adequate conditions, which can be well translated by the concept of Lorusso (1992), based on the definition of three well wooded areas, being, the public green areas, composed by the public patios destined for leisure or that opportune occasions of encounter and direct conviviality with nature. The private green areas, composed of the significant remaining vegetation incorporated into the urban network and afforestation of streets and public roads.

About the green areas that serve to shade Macapá, Bobrowskil (2011), Castro and Dias (2013), Sequeira (2014), warn that the selection of trees for pavements, streets, among others, is made by the residents themselves who choose to plant fruit trees or have tabular roots, which destroy pavements and generate conflicts with urban structures, besides not presenting a good leaf mass for thermal comfort, causing inconvenience to the homes and causing damage to companies that manage energy supply, water and public sanitation.

The damage includes participation in the city's local problems, influencing the statistics of Hougtonet al (1996), which warns that the global temperature of the earth has increased between 0.3 and 0.6°C since the end of the twentieth century (XX), as a result of the influence of natural and man-made elements, with humans being the product of human activity, generating

---

[123] Subject M.O.S., 28 years old, ordinary citizen, city of Macapá, Amapá, Brazil.
[124] Subject Z.N.M., 55 years old, professor at UNIFAP, city of Macapá, Amapá, Brazil.

Greenhouse Gases (GHGs) in the atmosphere, resulting from the burning of fossil fuels and changes in the pattern of use of common resources.

Here are some testimonials from Macapaenses about the effects of the absence of shading:

> [...] the city is destroying many things, I remember a few years ago that all that exists there, where today they are luxury neighborhoods, these buildings are not even ten years old; the change is very recent and should have started in 2010; it is difficult today to see areas of pasture and agricultural production, everything has become a city, the tree has become cement, the city is missing the shadows, it is in need [...]. (Verbal information)[125].
>
> [...] in the months of August, September and October, when it is very hot, we miss a lot of shadows; the birds have disappeared, the dust has increased, everyone has the flu and is breathing badly; when it is rainy season, the city is completely flooded, because it has no good sanitation structure, which to ensure or absorb water because there is no vegetation around the city [...]. (Verbal information). [126]
>
> [...] the afforestation project of this city sins for starting seedling planting projects only for new neighborhoods, leaving the downtown areas with nothing; a good proposal for shading the city happens in the city Curitiba I saw this, there is shade everywhere; the city is well cooled; people are polite and very cheerful, there are woods everywhere; in each corner only tourists visiting each forest. Finally, it is very different and pleasant to live in a city like that [...]. (Verbal information). [127]
>
> [...] I heard my colleagues talking badly only about the City Hall, nobody dared to talk about the seedlings placed by the landscaping division of the Municipal Department of Environment, which were all stolen in some points of the city, as in the roundabout of 13 de Setembro Avenue and Pedro Lazarino Avenue, among other points of the city; stealing public seedlings implies not knowing the damage it does to the individual as a citizen and to the city public in general. We really need to promote environmental education with emphasis on the importance that shadows and plants have for our city [...]. (Verbal information). [128]

To be pleasant to walk through finding ways to improve microclimates respecting the socio-environmental dynamics of the city of Macapá, as well as of Here, identifying and selecting compatible arboreal and bush species for the soil and regional climate, observing the[129] Foliar Area Index (IAF) for the formation of crowns that absorb or reflect solar radiation producing more shade, that is, the sunlight is more intercepted by its canopy, which will apparently be denser. The height of the tree determines the area or spatial dimension to be covered by the shadows in a given time interval. Shadows in the city are projected by pedestrians, vehicles,

---

[125] Subject V.A.M., 30 years old, ordinary citizen, city of Macapá, Amapá, Brazil.

[126] Subject L.S.M., 31 years old, ordinary citizen, city of Macapá, Amapá, Brazil.

[127] Subject A.K.N., uninformed age, SETUR technician, city of Macapá, Amapá, Brazil.

[128] Subject A.A.M., 34 years old, SETUR technician, city of Macapá, Amapá, Brazil.

[129] It is the ratio of leaf area to a given unit of land area, an important biophysical and structural parameter of vegetation that allows the interception of sunlight by the canopy of trees, creating shade (OLIVEIRA *et al*, 2013).

buildings, traffic signs among other objects and things, and their dimensions need to be compatible with the projections.

2.5   Shadow in agriculture

For Stafford-Smith *etal* (1985), Belchior (2014), shading through trees reduces about 55% of ultraviolet radiation and 85% of visible light contributing to decreased health problems by reducing body heat and facilitating regulation.

Beyond the term regulation shadows improve in agriculture, biological activity and soil fertility, especially if the tree is associated with microorganisms that fix nitrogen from the air, according to Carvalho *et al* (2002) and Linet *al* (2001).

Based in Belchior (2014), it can be said that both places of study have vegetables as their most prized diet, due to their rapid production in home gardens or not, despite being a plant typically from temperate regions.

In tropical regions such as the city of Macapá and the village of Aqui, it is recommended to grow in humid spaces of black or reddish clay soils with low water retention, generally located in river basins, to compensate for the low tolerance to solar heat, recommending shade cultivation for heat compensation.

A local of Aqui commented that "[...] in the case of lettuce, its cultivation under a shade prevents it from having a long and low quality stem [...]" (Verbal information)[130].

Shadows for agriculture are divided into two main types: natural, based on the use of shadows projected by trees larger than vegetables, requiring mastery of the shade path at the hottest time, while the artificial ones are formed by sunscreens.

According to the following statement in Macapá, the AgroAmapá Project was responsible for informing about the plantation in a controlled environment:

> [...] I have already participated in a lecture at the headquarters of Sebrae, dedicated to the cultivation of vegetables, grains, fruit and cassava, through the AgroAmapá Project, one of the purposes was to learn the techniques of planting in a controlled environment, with cover and suspended beds, this is very cool because the plant takes advantage of the shade to grow [...]. (Verbal information)[131].

According to Primavesi (2006a), shading improves the nutritional quality of short term pulses, when produced in regions with a warm equatorial climate, such as the city of Macapá, it is recommended to plant them under shade to take advantage of the three (3) or four (4) hours a day, which they need to gain nutrients and mineral salts.

---

[130] Subject J.A.C., 48 years old, village of Haqui in Massinga, Inhambane, Mozambique.
[131] Subject U.V.A., 27 years old, ordinary citizen, city of Macapá, Amapá, Brazil.

Fast-growing plants can be produced in temporary shade, designed from fast-growing plants such as banana and papaya trees, aiming at immediate protection or produced over permanent shade of trees and slow growth such as cashew, mafureira, mango, coconut, among others.

The following are the testimonies of the inhabitants of Aqui e Macapá who described their experiences with vegetable plantations.

> [...] we only planted tomatoes, lettuce, onions, cabbage in the machongo, but it turned out to be a borrowed space, so when the time for lettuce came, my wife tried to sow under the cashew tree; she watered every day and we produced lettuce stems that were larger than those below [...]. (Verbal information)[132].
>
> [...] at home we don't plant anything because it is all paved, everyone can hang some plant under a window, bedroom, somewhere that doesn't get sun and continues living, giving good ornamentation to the balcony, the living room, this practice is common there in the condominium [...]. (Verbal information)[133].
>
> [...] planting lettuce, tomatoes and scallions among other vegetables is common in the dry season, I have verified, when I go to visit my aunt in the hangover Vila dos Oliveiras which is in the south zone of the city; there one takes advantage of humid lands to plant, usually under a banana tree or another tree that supports excess water to help regulate humidity and mineral salts [...]. (Verbal information).[134].
>
> [...] when it's time for lettuce, we eat and what we sell at the market, the money is used to buy sugar and notebooks for the children to go to school [...]. (Verbal information)[135].

In the absence of trees, shading with fabrics is recommended where variations in their production depend on the artificial properties of the material that produces shadows (TABLE 1), so Primavesi (2006b) guides to the use of differentiated artificial fabrics (PHOTO 24).

Table 1 - Shading of low photoperiod cultivars

Treatment	Stand% of commercial plants	Weight of plants	Income	
			Stand Plant/ha	Kg/ha
Open skies	39,20	183,54	29.008	5.324
Silver screen	85,45	119,45	63.233	7.543
White screen	92,80	199,76	68.672	13.719
Black screen	83,50	158,99	61.790	9.824

Source: The author (2018).
Note: As of Epagri (2001).

---

[132] Subject A.H.T. 67 years old, village of Aqui in Massinga, Inhambane, Mozambique.

[133] S.M.M., 38 years old, ordinary citizen, city of Macapá, Amapá, Brazil.

[134] Subject L.S.M., 31 years old, ordinary citizen, city of Macapá, Amapá, Brazil.

[135] Subject A.H.T., 67 years old, ordinary citizen, city of Macapá, Amapá, Brazil.

Source: Jéssica Alves/G1 (2015).

The picture reflects the type of shading for the production quality of cultivars with low periodism, which constitute the food base of the inhabitants of both Macapá and the village of Aqui. It should be noted that when shading is with canvas, it becomes essential to choose the best material that allows qualified yields.

According to the inhabitants of Macapá and Here it was possible to notice in the vegetables better yield when the shadows were used:

> [...] I never asked anyone if the lettuce was shadowy or not, but what prevails is that its quality varies, I've noticed that. The variation is noticeable when looking, picking and eating, that I can classify [...]. (Verbal information)[136].
> [...] it's quite different, the feet of the one in the shadows are very big with a loaded green, the taste is very crunchy, not the machongo [...]. (Verbal information)[137].

In addition to agricultural production, shadows are ideal places for the conservation of vegetables, which even after harvesting maintain an accelerated metabolic activity, sun exposure even in short periods, accelerates dehydration and deterioration making them susceptible to putrefaction and less suitable for consumption, and the creation of Shadow Movable Units (UMS) is recommended for conservation.

Another advantage to prioritize natural shading is that it participates in carbon dioxide filtration, oxygen production, improves ecosystems, controls erosion, participates in microclimate

---

[136] Subject V.A.M., 30 years old, ordinary citizen, city of Macapá, Amapá, Brazil.
[137] Subject F.L.S., 69 years old, village of Aqui in Massinga, Inhambane, Mozambique.

change leaving micro and macrofauna favourable, helps to improve soil fertility, as well as increasing biodiversity, diversifying the income of the family producer.

For Head (1995), in intensive and extensive livestock farming shadows ensure welfare during periods of high incidence of sunlight, decreasing the temperature of the environment and influencing the decrease in the temperature of the animal underneath it, causing increased food and water intake to the animal including environmental benefits (PHOTO 25).

Photo 251- Shade tree for animals - Haqui village

Source: The author (2015).

The photo was taken in February 2016, at about eleven thirty minutes in the morning, (11:30), at the moment when the sun is at its height. According to Fraser and Broom (1990), Conceição (2008), the cattle perform four basic activities, which are displacement, grazing, rumination and laziness, and when the cattle are comfortable, without any kind of stress, they perform the activities of laziness and rumination in the lying down position, and the standing position indicates stress.

> [...] here in the village, oxen have been extinct, the few families that have some use the technique of tying the animal under the shade of a tree, just as we do with goats, donkeys and pigs. There are two families in the eastern zone that still use the grazing system accompanied by a boy who serves as a shepherd. He takes the animals to the plains which are no longer plains because everything is busy. At the hottest time the oxen don't even need to be oriented, they themselves look for a good shade to rest and finish eating, all the grass they swallowed during the cold hours [...]. (Verbal information)[138].

---

[138] Subject J.A.C., 48 years old, village of Aqui in Massinga, Inhambane, Mozambique.

> [...] ducks and chickens do not mix at home, each group of animals already
> knows their shade at noon; if ducks invade the chickens' shade there is war; in
> the hot hour all the animals like to rest until they sleep, the chickens even dig to
> take advantage of the humidity in the shade [...]. (Verbal information)[139].

Barion (2012) recommends the use of artificial shading by means of screens, in situations of absence of trees in the pasture or while waiting for them to grow. Resident of Macapá confirms with Barion that there are organized shading for livestock:

> [...] I am not an expert in livestock farming, but as far as I know, there is an
> intensive and extensive system of breeding in shading of cattle, muare[140], birds
> and buffaloes within our state as is the case in the municipality of Santana [...].
> (Verbal information)[141].

Both the animals raised in one system and the other need shade to maintain the homeostasis or thermal balance of the body, and in Macapá the intensive system practiced by farmers prevails, already in the village of Aqui, the extensive over the tropic of capricorn.

---

[139] Subject F.L.S., 69 years old, village of Aqui in Massinga, Inhambane, Mozambique.
[140] Set of mules.
[141] Subject J.A.M., 40 years old, city of Macapá, Amapá, Brazil.

CHAPTER III

TROPIC OF CAPRICORN AND EQUATORIAL LINE

This chapter discusses the categories, tropic of capricorn and equator, as imaginary lines that cross the globe in parallel, forming part of the five main circles of latitude that delimit the main climatic regions of the earth's surface, with solstices and equinoxes occurring respectively. Capricorn crosses ten countries, already the line of the equator thirteen, influencing diverse experiences of the peoples, who use the phytogeographic characteristics influenced by the passage of these two imaginary lines through their territories. Throughout history, the tropic concept has been used to designate the colonized territories and peoples in Africa, Asia and America, other times to the movement of claim to Eurocentrism as happened in Brazil with tropicalism and tropicália. In the geographical approach, the word tropic symbolizes the solar position and angle of solar slope, responsible for the environmental dynamics and the thermo-pluviometric variations, distribution of fauna and flora, as well as the behaviour of living beings, contributing to the development of adaptation mechanisms of living beings in line with photoperiodism, foliar abscission and homeostasis. The research work arose in the context of Teaching and learning Geography, seeking to explain the influence that imaginary lines play in the experience of the residents of the city of Macapá in Brazil and the Povoado de Aqui in Mozambique, both crossed by the line of the capricorn tropic and the equator.

3.1 Tropic of capricorn and the zero shadow (solstice)

About Capricorn, Alves (2006), Afonso (2006), Cherman and Vieira (2011) and Rodrigues Júnior (2012) are unanimous in agreeing that, in order to study about epistemology, it is important to look for a combined basis, anthropological and geographic studies that consider the tropics as regions where people live and biodiversity that tolerates the climatic variations of the regions that extend from twenty-three (23°) positive degrees in the Cancer in the North to twenty-three (-23)°negative degrees in the Capricorn in the South.

Among the scholars of the subject, Erastóstenes (about 5th century a.n.e.) is pointed out as a pioneer to approach about the limits of the solar trajectory, when he found on papyrus in the library of Alexandria, an information announcing that in the city of Siene, present-day Aswan in Egypt, at noon there was a summer solstice around July 21. The sun was at ninety (90) degrees, illuminating the deep waters of the well, without causing a shadow, deducing the apparent solar circularity.

The intensification of debates about the tropics gained new momentum when astronomy studies succeeded in establishing precisely the coordinates of the solar ecliptic at 23° south and north (N/S) symbolizing the "[...] apparent solar movement from one solstice to the other".

(CONTI, 2010, p.49). Inspiring Eurocentric positions, in some geographers like Emanuel de Martonne, 1946 and Gourou, 1948, who advocated the use of the term only for hot and humid areas. Demangeot ([s. d.]), Planhol and Regnon (1970), extended the concept to semi-arid areas, serving as criteria for demarcating desert regions and later the sad tropics of Claud Levis Strauss in 1955.

The importance of the tropics continued to motivate studies, and movements contrary to the Eurocentric current emerged in the USA and Brazil; in 1960, the "[...] tropicology of Gilberto Freyre and the tropicalism of Edson Fonseca [...]" (FAVARETTO, 1996, p. 2). Movements that demonstrated that the humid tropics were not wild were geographical spaces with their own phytogeographic characteristics and peoples with capacities identical to those that inhabited European cities and towns. The manifestations of this movement were made in various forms, such as art, poetry, lyrics, music, etc., mainly in the great Brazilian urban centres of the time, such as São Paulo, Rio de Janeiro, Recife and Pernambuco.

Conti (2010), allows us to say that this movement, opened space for a new approach to the tropics that came to be understood as a category of analysis that aggregates a multiplicity of environmental variables, which includes cultural aspects addressed in the anthropology of the "sad tropics" of ClaudLevis-Strauss, the historical and geopolitical elements with emphasis on the colonial moment that incorporated the concept to outline projects of the productive system of medium latitudes, producing crops such as cotton, tea, rubber, coffee, etc. With these studies, the tropic becomes understood as a climatic belt with its own fauna and flora, as well as with diverse peoples who live and make the region their wealth of exploitation of resources for survival. In this perspective, the colonial system recognized the region's potential for the production of export crops in the international market of the time.

The concept is currently used as a theoretical framework to establish solar apparent displacement limits in the southernmost or northernmost position, limiting the maximum incidence of solar rays on the earth's surface, marking the transition from warm to tropical equatorial climate in the southern or northern hemisphere.

Based on authors such as Benchimol (1990), Silva (2006), Chermane Vieira (2011) and Varella (2013) it is possible to state that the tropic of capricorn is an imaginary line demarcated by the coordinate 23° 26' 22" south latitude, variable depending on the author, for Silva (2004) is 23° 27', Carvalho Junior *etal* (2015), 23,4378° South (23° 26' 16"), with others considering the tropic line located at 23° 30'; 23° 45', onwards, occupying a strip of approximately nine hundred and eleven point three (911.3 Km) kilometres and one hundred and two point two (102.188 Km) kilometres long.

The basis for determining the moment when the sun reaches the solstice position is from the solar declination (δ), which is the angle formed between the earth's equator and the imaginary line containing the plane of the sun. Calculations can be made on the basis of different mathematical formulas, namely: $\delta= 23.45 * sen\ [360(248+DJ)/365]$; $\delta= 23.45 * sen\ [(360/365).(DJ - 80)]$ or $\delta= 23.45 * sen\ [(2.(\pi)/365).(284+DJ)]$; where δ is solar declination; * a Multiplication; Sen o Seno; DJ to Julian Day and Π to the sixth decimal letter of the Greek alphabet Pi.

The capricorn line aggregates in its theoretical dimension geographic spaces with seven spatial dimensions (latitude, longitude, altitude, temporalities, subjects, objects and things), of the ten countries crossed namely Australia, Madagascar, Namibia, Zambia, South Africa, Mozambique, Brazil, Argentina, Chile and Paraguay (MAPA 4).

In Mozambican territory the capricorn tropic line cuts through Inhambane province through the Massinga district in the village of Aqui, the eastern part of the Panda district and crosses Gaza province through the Chigubo and Chicualacua district until it enters the Limpopo province in South Africa. In this context, the Tropic of Capricorn is no longer just an imaginary line, but becomes a three-dimensional socio-environmental geographical space, as it is perceived, conceived and lived (LEFEBVRE, 1974, 2006). In this case, Capricorn, symbolizes, geosystems that aggregate networks, lines, webs, ontogenesis, autopoese, among other forms of referencing the systemic complexity, being constituted these places by environmental variables represented by air, soil, water and living beings, which form a totality called geodiversity.

Geodiversity, allows the inhabitants of the places crossed by the capricorn line, to maintain their own *modus vivendus,* which differentiates them and resembles the social experiences of other peoples situated at the same latitude. The differences arise in the struggle for survival, where each social group finds different ways to adapt to the phytogeographic conditions, provided by the geographical situation, in a region where the sun makes the maximum decline at the moment of its apparent movement.

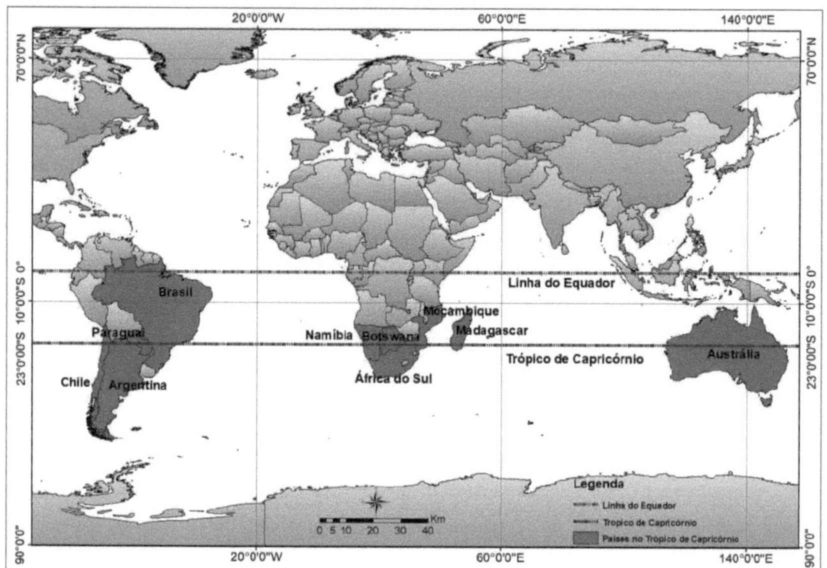

Source: The author (2018).
Note: Using HQIS 2.18.15 *las palmas*.

Since the sun is the primary source that emits about ninety-nine point seven (99.7 percent) of energy on the earth's surface, its distribution depends on its declination or position in relation to each place in the geographical space, and for the village of Aqui, the sun is at a declination of -23.45 south, see Table 2.

Table 2 - Solar position in the village of Aqui

DAY OF THE MONTH	YEAR'S DAY NUMBER (NDA)	SOLAR DECLINATION (δ)	NOTE
15th January	15	-21,27	
15 February	46	-13,29	
15 March	74	- 2,82	
21 March	80	0,00	Water Equinox in Macapá
15 April	105	9,41	
15 May	135	18,79	
15th June	166	23,31	
21 June	172	23,45	Winter solstice
15 July	196	21,52	
15 August	227	13,78	
15 September	258	2,22	
20th September	263	0,00	Macapá drought equinox
15 Otubro	288	-9,60	
15 November	319	-19,15	
15 December	349	-23,34	
20 December	354	-23,45	Shadow Festival here

Source: The author (2017).

The declination was obtained from calculations based on the following mathematical formula, $\delta= 23.45 * sen [360(248+DJ)/365]$, and on January 1st the sun is positioned south of the equator with a declination of (-23.01) = at 30 (-17.78) and on February 28th with a declination (-8.67); during the days (31) March, (30) April, (31) May, (30) June, (31) July and (31) August, the sun is in the northern hemisphere with a magnetic declination of (+3.62), (+14.59), 31 (+21.90), (23.18) and (+18.17), +8.10) respectively. The summer solstice in the village of Aqui in Massinga takes place on 21 December, at the moment when the sun reaches a magnetic declination of -23.45 = to -27" = -0.45°.

The NDA corresponds to the Julian Day Number (DJ) which represents the day of the year independent of the months, that is to say, it represents the sum of days, from the first of January until the date that you want to calculate the solar declination, for the case of this study, the day of the year allows to calculate the position of the sun on any day of the year in relation to Macapá and Here. In the city of Macapá, when the declination is zero, the water equinoxes take place in March and the droughts in September, while in the town of Aqui, when the sun is at its highest declination, the summer solstice is recorded in December, known by the locals as the time of the December fruit premisses or shadow festival.

While scientifically the sun's apparent mobility is calculated from mathematical formulae, the residents of the city of Macapá and the village of Aqui in Massinga use their everyday knowledge and practices, which from the window of their flat or dwelling analyse and explain the variations of solar apparent movement throughout the year.

According to GundoImbrie (1979) quoted by Silva (2007), Milankovitch's mathematical discoveries better explain the variation in the intensity of the effects of insolation with latitude, referring to the influence of the cycle of obliquity, the inclination of the Earth's axis and the cycle of precession that cause changes in the date, time and day of equinoxes and solstices (TABLE 3 and 4).

Table 3 - UTC[142] date and time solstices between 2014 and 2025

Year	December solstice	
	Day	Time
2014	21	23:03
2015	22	04:48
2016	21	10:44
2017	21	16:28
2018	21	22:23
2019	22	04:19
2020	21	10:02
2021	21	15:59
2022	21	21:48
2023	22	03:27
2024	21	09:21
2025	21	15:03

Source: The author (2018).
Note: From NASA data, 2010.

Table 4 - Equinoxes between 2014 and 2025

Year	Equinox March		Equinox September	
	Day	Time	Day	Time
2014	20	23:03	23	02:29
2015	20	04:48	23	08:21
2016	20	10:44	22	14:21
2017	20	16:28	22	20:02
2018	20	22:23	23	01:54
2019	20	04:19	23	07:50
2020	20	10:02	22	13:31
2021	20	15:59	22	19:21
2022	20	21:48	23	0:40
2023	20	03:27	23	06:50
2024	20	09:21	22	12:44
2025	20	15:03	22	18:19

Source: The author (2018).
Note: From NASA data, 2010.

The U.S. Space Agency, which is responsible for research, technology development and space exploration programs, published in 2010 a list of data indicating the universal day and time when solstices and equinoxes will occur by the year 2025, facilitating research and observations of these phenomena.

---

[142] Coordinated universal time.

In each four-year cycle the equinoxes tend to be delayed, that is, over the same century they tend to happen earlier, due to the orbit of the earth that runs faster when it is closer to the sun or perihelion than when it is further away or aphids. The behaviour concerning the variation of solar radiation in the village of Aqui can be represented by means of the following Chart 5.

Chart 5 - Variation of solar radiation here

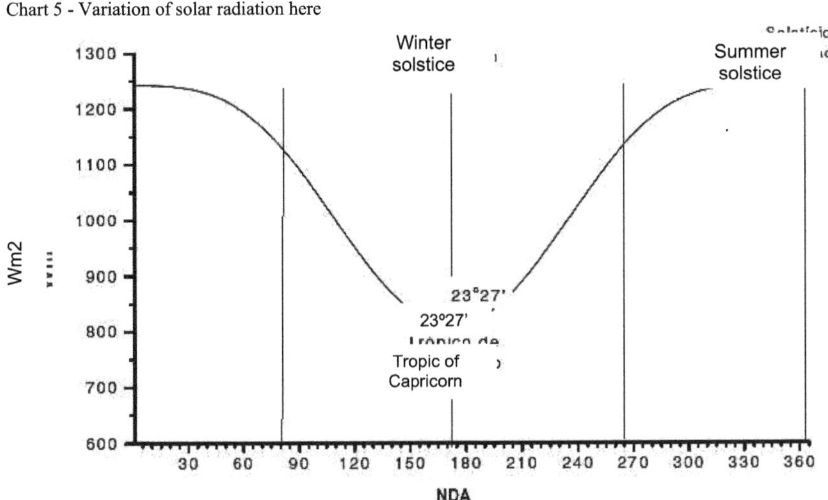

Source: The author (2018).
Note: From Silva (2004).

Solar radiation varies with latitude throughout the year, and for the village of Aqui, situated on the imaginary line of the Tropic of Capricorn on the parallel of 23°26'22" south latitude, will receive the greatest amount of solar radiation all year round, reaching higher values at the time of the summer solstice, about (1250 W/m²), receiving below eight hundred (800 W/m²) watts in each unit of time in one meter quarter during the winter solstice. These variations make the village responsible for the typical local socio-environmental diversity and with biodiversity variations capable of supporting these changes throughout the year by mobilizing socio-environmental organizational structures.

Organizing social and environmental structures means that the animals and people that inhabit this geographical area of planet Earth, organize their daily lives to the detriment of thermal variations by establishing a photoperiod. Animals, plants including humans, have a homeostasis or standard temperature, known as photoperiodism in plants, requiring adaptation from temperature-regulating strategies in the body of living beings. In this perspective, Macapá residents use air-conditioning, beach baths, shading, other forms of ventilation and refrigeration

93

of spaces or living environment. In the village of Aqui, they use shading of trees, of huts, tents and other objects that soften high temperatures.

Animals and plants that do not respond in optimal spectrum, such as the camelão, the leseira baré, known as laziness of Bahia, walk or move slowly to reduce basal metabolism and allow the body to continue functioning. Scientifically, the photopriod can be calculated in such a way that plants and animals can withstand outside the homeostasis situation. The mathematical formula is, $F=\{2/15*arc.cos[-(tg\varphi*tg)+1/60[43,7864+0,15150+0,01330]$, where, $+1/60[43,7864+0,15150+0,01330]\}$ corresponds to the twilight which is equal to $\pm 0,73$.

3.2 Line of Equator and the Equinoxes

For Rodrigues Júnior (2012), Araújo (2014), Trogello (2013), Ecuador is an imaginary line of the earth's surface in which the astronomical latitude is equal to $00° 00'00"$ dividing the earth into two hemispheres "[...] geographic north and geographic south [...]" (MILONE, 2003, p. 25). With a radius of 6 378 km corresponding to the perimeter of 40 075 km. Cutting through or crossing theoretically three (3) oceans (Pacific, Atlantic and Indian), four (4) continents (America, Africa, Asia and Oceania), thirteen (13) countries (São Tomé, Gabon, Democratic Republic of Congo, Congo, Uganda, Kenya, Somalia, Maldives, Indonesia, Kiribati, Ecuador, Colombia and Brazil), Map 4.

Map 4 - Countries crossed by the imaginary line of Ecuador

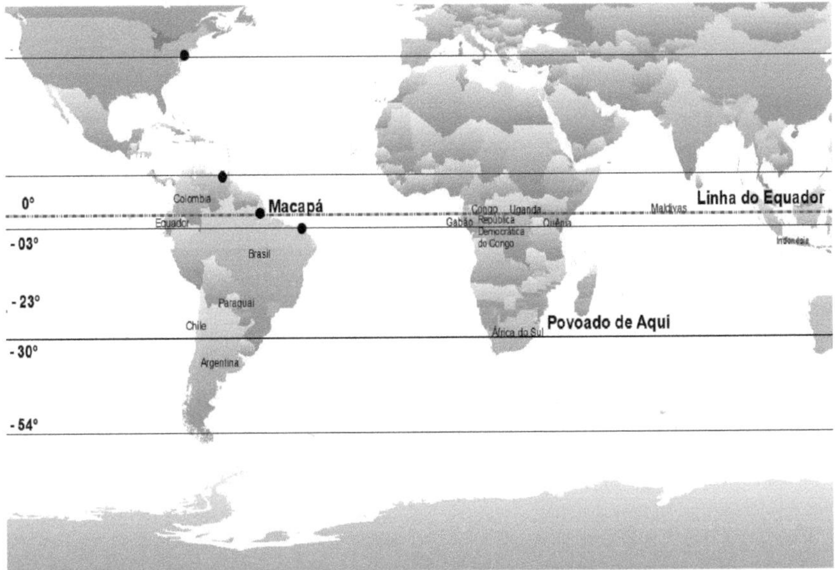

Source: The author (2018).
Note: Using HQIS 2.18.15 *las palmas*.

In Brazilian territory the line of the Equator crosses the state of Pará, Rorainópolis, Amazonas (São Gabriel da Cachoeira) and the state of Amapá cutting the city of Macapá. Ecuador represents a geographical area made up of territories populated by populations with different cultures, despite sharing the same latitude. These differences are notable for the diverse ways in which these peoples exploit the potential of the geographical situation, as is the case with the city of Macapá, which organises its political and socio-economic structures by combining its strategic question under the imaginary line of Ecuador, where the variations in the annual seasons are less pronounced.

According to Alves et. al. (2009), the sun in its earth translation movement crosses the imaginary line of the equator twice a year, making day and night have the same duration all over the planet. In the instant the sun is at the zenith. Although the equinoxes happen only twice a year, the variation of solar radiation in the countries crossed by the equator does not vary throughout the year (CHART 6).

Graph 6 - Variation of solar radiation in Macapá city

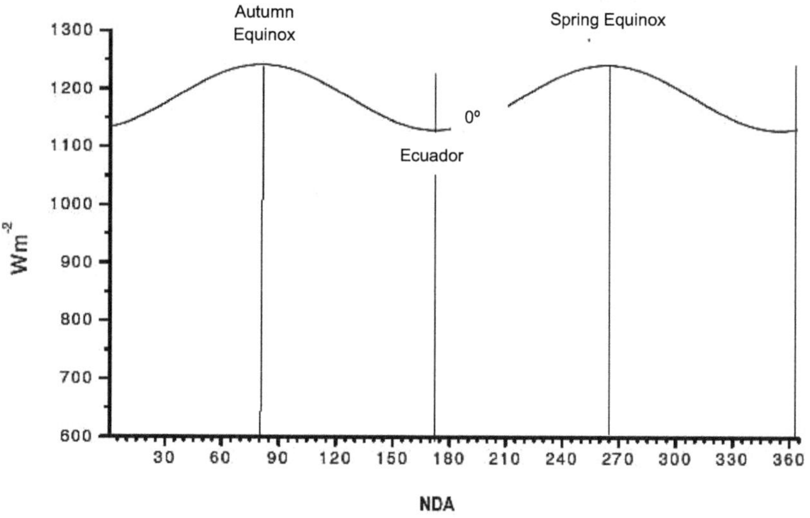

Source: The author
Note: From Silva (2004).

In the city of Macapá the intensity of solar radiation is high throughout the year, reaching extreme values during the equinoxes of the droughts, on twenty-two (22) September and the waters, on twenty-one (21) March.

The position of Macapá allows the sun's rays to travel a smaller distance in relation to the horizon plane, making the place receive a high amount of solar radiation, while the village of Aqui, because it is situated at medium latitude, which is further away from the equator, will receive a relatively smaller amount when compared to Macapá. This is because at the height of the sun over the horizon, it reduces and the sun's rays travel a longer distance in the atmosphere before reaching the surface of the village.

3.3   Macapá and Haqui Shadow Diagraming

Starting from Frota and Schiffer (1995), Corbella and Yannas (2003) it is possible to affirm that the solar diagrams or charts are projected representations of the trajectory or itinerary of the sun on dates referring to the solstices, equinoxes or intermediate dates. Facilitating the analysis and interpretation of sunshine and the position of the sun on a given date of the year at each point on the earth's surface, as well as the determination of the shading projection by objects and things for ventilation of a given place. The solar chart is a graphic or schematic representation

96

demonstrating the path the sun travels in the sky during the day, influenced mainly by various natural and anthropic factors.

Ribeiro (2003) explains that solar diagramming depends on the position of the observer corresponding to the azimuth and height in relation to the sun, requiring the calibration of the solar clocks to the detriment of the geographical coordinate, this means that it is important to identify the geographical quadrant during the diagramming. If the diagram is carried further north or south the lines of its quadrant no longer correspond to the hours of the day of the place, because the sunshine and shading have altered its trajectories and characteristics.

Knowing that the position of the sun in Macapá is different from the village of Aqui, it is important to make an analysis in order to compare the experiences of use and shade use of the inhabitants of the two geographical areas of study, so we present the diagram (GRAPH 7).

Graph 7 - Macapá solar map

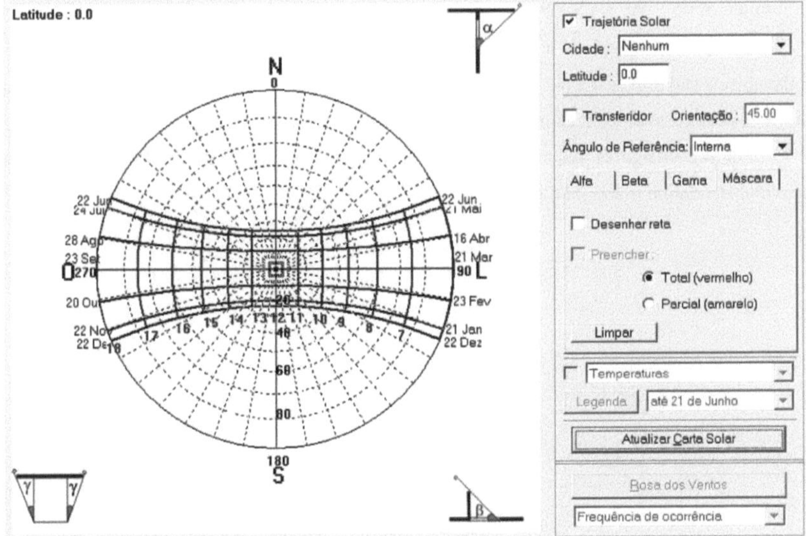

Source: The author (2017).
Note: Using air-to-air software 6.2.

With the use of the *6.2 air softweresol, it* was possible to represent a stereographic projection of the celestial vault, in a horizontal plane symbolizing the various positions of the sun and shadows throughout the year in the city of Macapá. The horizontal and vertical curved lines represent the months and hours of the day respectively. The horizontal lines indicate the projection of the solar path during the months of the year, the hours of the day are represented by the vertical lines (6 am to 6 pm), showing the azimuth angles and the solar height projected in one plane.

The Macapá solar map contains three sources of information: an external ring with angles to the north (N); a mesh that facilitates the location of dates and times; and a tracking system at the bottom that provides the inclination of the rays (h), that is, the horizontal lines represent the dates of the year, and the vertical lines represent the times of the day to know in which position the sun will be on that date, and can project the position of the shadows. The projected shadows are at an angle of ninety degrees (90°) during the local midday. Macapá, has a solar radiation above thirteen hours daytime during the year, shadows projected to the west in the morning and to the east at the end of the day, always overlapping about half a local day.

To calculate the Vertical Angle of Shadow (AVS) projected in the city, it can be calculated taking into account that the sun is at an inclination of ninety degrees (90°), and using the formula, AVS = 90, AVS = 90°- 0°; AVS = 90°. That is, the Vertical Angle of Shadow by the city of Macapá corresponds to a slope of ninety degrees (90°), that is, the shadows are projected by the objects and things at a perpendicular slope corresponding to an overlap throughout the year with little projection to the East in the mornings and West in the afternoons.

For the village of Aqui, which is situated in the middle latitudes will show another behaviour in relation to the shadows (GRAPH 8).

Graph 8 - Solar map of the village of Aqui - Latitude, 23°26'22"

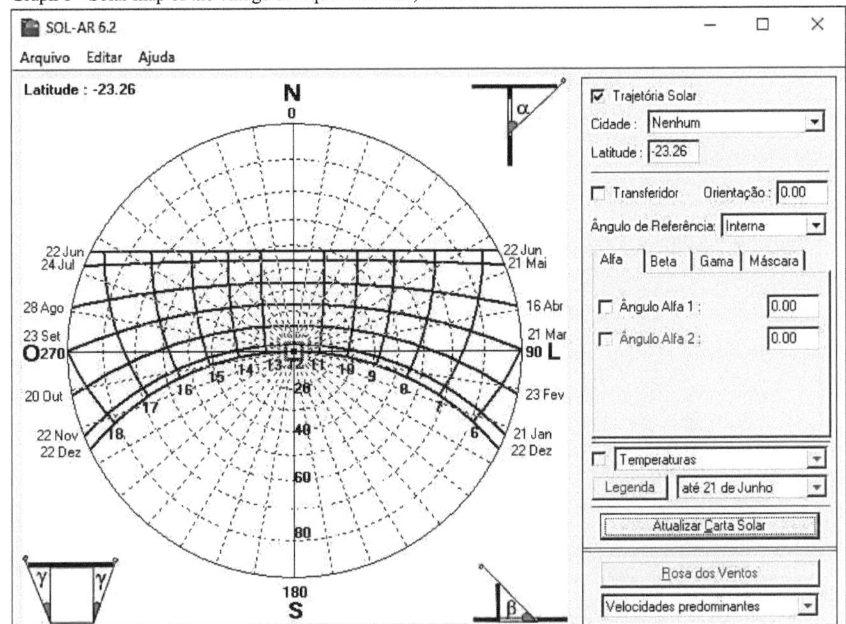

Source: Prepared by Carlitos Sitoie (2017) using *Softwaresol-ar 6.2*.

The solar diagram shows the different routes or points of the sun and shade in the village of Aqui throughout the year. The horizontal lines are on average 30 days representing the months and hours of the day respectively. The radial lines indicate the azimuth or bussolar position in relation to the magnetic north, while the concentric circles indicate the altitude of the sun from sunrise to sunset. In other words, the vertical lines that indicate the (5h30 to 18h30) indicate the position in which the sun will be at that date in the settlement, allowing to predict the position of shadows and sunlight. The angle of solar slope can be calculated using the AVS formula = 90°-23°26"22'. The sunshine or insolation exceeds an average of fourteen hours and the shadows are projected indicating three positions East, Northwest and Southeast. The sun has a slope ranging from ninety (90°) degrees to twenty-three (23)°degrees.

CHAPTER IV

SOCIAL EXPERIENCES THROUGH THE SHADOWS

The chapter deals with shadows as physical phenomena resulting from the absence of light. Their extent depends on the height and volume of the object blocking the light. It makes these phenomena important school subjects with emphasis on trigonometry, which makes it possible to determine the position and size of shadows projected over time and geographical space. Knowledge about shadow projection is useful in civil construction through thermal comfort planning and solar diagramming, in agriculture and cattle breeding for the regulation of photoperiodism of agricultural crops and homeostasis of animals, as well as to plan quantities of males and females in oviparous, among others. The shadows most used in the village of Aqui and in the city of Macapá are projected by huts, tents, buildings and masonry houses, coconut trees, ipês, jambeiros, cashew trees, hoses, among other tree species of wide canopy used for thermal comfort. The results of the research indicate that shadows in the city of Macapá are projected out of the reach of the users causing thermal discomfort, leading the residents to go to public squares at night, walk outside the pedestrian pavement, looking for shadows in road lanes. In the town of Macapá, there is concern in the allocation of objects in order to offer shadows in places of social conviviality in the hottest hours of the day, because shadows mean hiking trails, places of living impregnated with daily life.

4.1 Shadows and the sun

According to Silveira and Axt (2007), shadow is a subject of study in optics that considers the dark region formed by the partial or complete absence of light, resulting from the obstruction of illumination by an object that serves as an obstacle, making it difficult to penetrate luminosity. In this perspective, shadow is the dark part that occupies the back of an object and may change position according to the source of light. For physicists, shadow is something that does not exist, because it symbolizes the absence of light, and neither are the objects that project the shadow, but rather, it is the reflection of the lack of light.

According to Silva (2006), the capacity of objects to block light is evaluated according to their opacity, which determines the quality of the shadow, being either clear (it does not let the light pass) or partial that which allows the passage of light. As for extension or size, the shadow depends on the distance between the object and the source, the volume and the height of the blocking object. Besides being clear and partial, shadow as a physical phenomenon can also be its own, that which is formed by the object itself, by the effect of the incidence of light on the object. Projected, when an object in contact with light forms a shadow that is projected later on a plane or even on another object.

Still based on Silva, it can be said that the quantification of a shadow area depends on the shape and dimensions of the object that produces it, including the angle of radiation incidence. For the purpose of determining the shape and position of a shadow, the dimensions of the tree and the approximate geometric shape of the object or thing that projects the shadow are necessary. For the study the following basic formats were considered:

Hexagonal-shaped objects were considered to be haystacks and tents because of their irregular shape resembling geometric figures with six (6) sides belonging to the group of hexagons (FIGURE 2 and 3).

Figure 1 - Shadow projected by haystacks from Haqui village

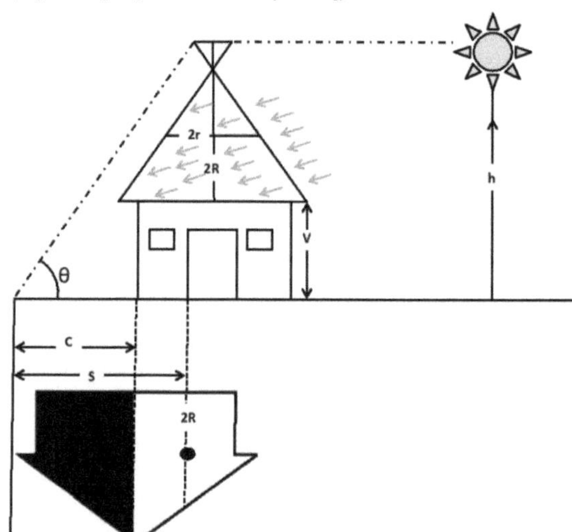

Source: Marques (2017).

Figure 2 - Shadow projected by Haqui tents

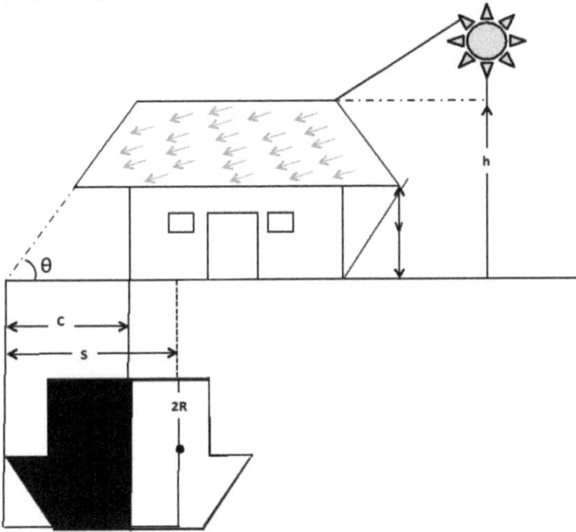

Source: Marques (2017).

The settlement's residences are built in a circular form (haystacks and barns), sometimes in various shapes that resemble the trapezium and hexagons, known locally as differentiated-style tents, which can be of reed and straw, cement and zinc or combinations of these materials, constituting the majority of the housing type built in the settlement. These constructions have several functions, being used for sleeping, but also for heat protection through the projected shade.

Other objects used for shadows in the settlement are of the same shape as cylinders such as the coconut tree (FIGURE 3).

Figure 3 - Cylindrical-shaped objects

Source: Marques (2017).

Most of the vegetation in the village of Aqui is made up of coconut trees, which are fed from the extraction of thatch sap and coconut derivatives, making it one of the village's main sources of income. Despite its abundance and food importance - commercial - the coconut tree is not a good source of shade, due to its narrow and elongated leaves projecting partial shade, associating it with its fruits, which when dropped on a person can cause injury, do not constitute the shadows preferred by the inhabitants.

Objects of inverted cone shape were identified, which constitute the canopy of ornamental trees and trees of streets, avenues, public squares of the city of Macapá or planted in the backyards, fields and paths of the village here. Among the main species, the ipê or *handroanthusalbusjamboeiro* or *syzygium*, cashew trees or *anacardiumorcidentale*, mango trees or *mangifera indica* stand out, among other species of broad canopy that serve as shade for the inhabitants of the city of Macapá and Haqui (FIGURE 4).

Figure 4 Hose shadow

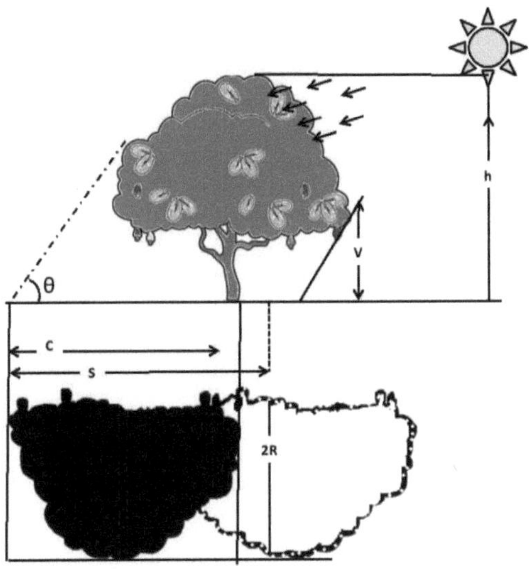

Source: Marques (2017).
Note: Adaptation.

Due to the shape of the conical and wide crown, known as the semicircular, they are generally preferred because they provide dense, opaque or closed shade, sometimes the trunk of the tree projects shade that reinforces the ventilation of the place, besides producing fruit that when it falls cannot hurt the individual, if compared to the fall of a coconut on the head of a person.

In the city of Macapá, rectangular objects that project shadows have also been identified, such as buildings built in various condominiums in the city of Macapá. The city has many condominiums formed by buildings and flats that project shadows causing solar obstruction in addition to blocking the good circulation of air masses, making some places in the city dark, as mentioned in some testimonies of the social subjects below (FIGURE 5).

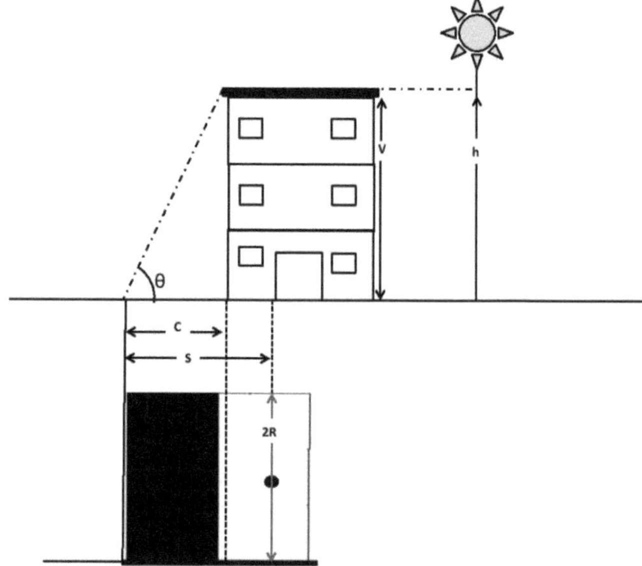

Source: Marques (2017).
Note: Adaptation.

In all figures the letter (r) means the radius of the roof and wall of the hut, tent, building or treetop, the letter (c) the length of the shade, the letter (s) the greatest distance of the shade from the wall, roof of the hut, tent, building or treetop. Thus, the smaller the azenithal angle (θ) the longer the hexagonal, ellipsoidal shape of the cone will be, at the end depending on the shape of the object that projects the shadow onto the ground.

The values corresponding to the length of the projected shadow, the displacement of the shadow in relation to the object or thing that projects area or surface occupied by the shadow on the ground can be calculated according to mathematical formulas of area, length, height, and sharpness, applied to teaching programmatic contents in the didactic unit that studies the geometric figures.

The following statements highlight the relationship between the planning of buildings and the shape and arrangement of objects that project shadows:

> [...] the construction of buildings in the city has become a fever, but there is one, but some condominiums have the buildings so piled up that not even a bit of sunlight enters. Everything is dark, almost all the time, the children and the

elderly who don't work are left without getting sun [...]. (Verbal information)[143].

[...] when planting a tree you should know where the shade goes, see the distance so as not to be out of the yard or on the way. But you should avoid that the house has been under the shade all day, this can make the house very uncomfortable especially in cold weather [...]. (Verbal information)[144].

In both cases, in the testimonies of Macapá and Haqui, one can notice the concern about the position and shape of the shadow, which when badly planned can cause discomfort, because it is also evident the importance of light and sun rays in the dwellings. Thus, the planning, both of houses and tree plantations in housing areas, public squares, walkways and hiking trails, must follow some environmental planning through shadows in order to provide thermal comfort and health for people and their animals.

Some experiences lived by the inhabitants of Macapá and the village of Aqui will be presented below, about the use of shadows for geometric orientation, the determination of calendars and timetables, demarcation of distances among other uses of shadows.

4.2 Geographic orientation and time determination through shadow

According to Almeida (2006) and Silva *et al* (2010), orienting means looking for the geographical orientation, which allows navigation from fixed references of the earth's surface, combined with the observation of the stars, requiring determination of the local North-South (N/S) meridian line of the observer. When navigation is based on shadows, it is crucial to determine the angle of observation or the coordinate of the place where the individual is positioned in relation to the incidence of the sun's rays "[...] for times varying between 0 and 24 hours depending on latitude and time [...]" (LIGHT, 2015, [s. p.]). The observer must have mastery of the references of the place, know the position in which the sun rises and sets, including the trajectory that the shadows make up throughout the day.

From Feteris and Hutton (2000), Jackson (2004) can identify the orientation method from the observation of solstices and equinoxes, positioning himself in a Fixed Observation Point (PFO). This means that by establishing the PFO, for example, there is a need to orient at any point in Macapá and Here, the procedure would be the same, with the rising sun at the water and drought equinoxes taking place in the centre of the eastern strip. Already in the winter solstice and the winter solstice rises at the left and right end respectively (FIGURE 6).

---

[143] Subject V.A.M., 30 years old, ordinary citizen, city of Macapá, Amapá, Brazil.
[144] Subject F.L.S., 69 years old, village of Aqui in Massinga, Inhambane, Mozambique.

Figure 6 - Guidance on the basis of solstices and equinoxes

Source: The orientation... (19--?).[145]

For Jackson (2004) Machado (2013), even knowing that the sun rises every day in the east and sets in the west, associated with the method of orienting himself geographically by means of equinoxes and solstices, the method is not very effective, due to subjectivity in determining the solar spring and interference from atmospheric elements, topographical irregularities, vegetation characteristics, constructions that can make it difficult to determine the fixed point of observation of the solar spring, apine[146] and sunset.

Despite the inefficiency caused by subjectivity in determining the sunrise and atmospheric phenomena in line with the superimposition of shadows, considered a moment of equinoxes and solstices in this book, the inhabitants of Macapá and the village of Aqui, learn small to mark a point on the ground, or put a stick to signal the end of their shadows or shadow projector objects, to determine the time for various day-to-day activities. The technique is empirical, based on the daily observation of the size and position of the shadows to mark the time for an activity, which can be school, leisure, etc. One of the examples is mentioned in the following R.N.S. speech:

> [...] when I was a hunter, I learned to move without knowing the direction in which the sun rose. All the clouds do not close the sun is to look for a place where the trees do not cover the sun's rays. But you must have the patience to measure our shade in a little while, marking every moment. Then, draw a line that passes through the various points marked and at the end positional in the central mark, facing the point on the line that is closest to discover the dzonga

---

[145] According to Portal Silvestre. Available at: http://www.silvestre.eng.br/astronomia/criancas/orientasol/.
[146]Zenith, the highest moment of the sun, the moment the ground is apparently over our heads.

> (North), from there it is easy to know autchene (source), aupelagambo (west) and uronga (South) [...].

In addition to determination on the basis of equinoxes and solstices, another method is based on the technique of a stem, which in addition to giving directions and azimuth for geographical orientation, determines the time through the shade, it being necessary to mark the shade two to three times of the day, which can be in the morning at any time, at noon and in the afternoon, then draw a circle joining the shadows, which will allow to trace a rose of the winds and the weather (FIGURE 7).

Figure 7 - Solar clock

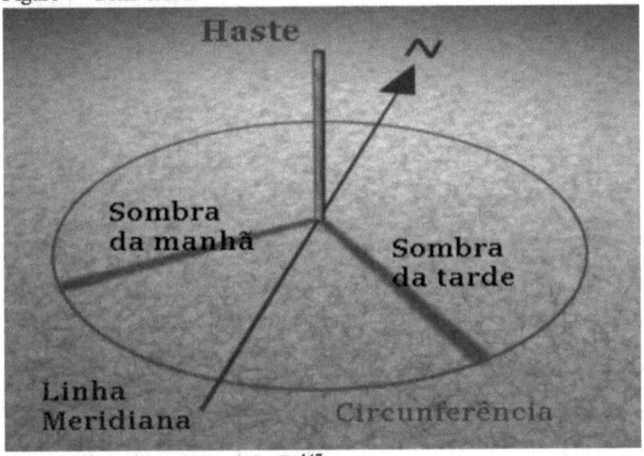

Source: The orientation... (19--?).[147]

The figure is a demonstration of the shadow recorded in the morning and afternoon, the circle in red, represents the length of the shadows projected in the two moments of the day, the Meridian line, drawn in blue, represents the bisector of the angle between the shadows, pointing north on one side and south on the other.

The mobility of the shadows happens in the opposite direction to the apparent movement of the sun, that is, while the sun rises in the east and sets in the west, the shadows perform the movement from west to east. When the sun rises, there are longer shadows of the day, projected to the west and shrinking until they become shorter at about noon local time, and increasing again to a greater length when projected to the east at the moment the sun sets in the west.

As the testimonies complement: "[...] what is easy to know is that the shadow in the morning is in the west in the afternoon close to us and at the end of the day in the east, that even a child of the first grade notes [...]" (verbal information)148 . "[...] the shadows are very big in the morning and small at twelve and they vote to be very big in the afternoon, then they disappear,

[147] According to Portal Silvestre. Available at: http://www.silvestre.eng.br/astronomia/criancas/orientasol/.
[148] Subject R.N.S., 55 years old, ordinary citizen, city of Macapá, Amapá, Brazil.

only stay at night, and then tomorrow they do the same thing again every day [...]" (verbal information)149

Orientation also requires adding knowledge about geographical references of the place, how to record or fix images of the landscape in the mind in order to "[...] locate, know, go to their destinations and recognize when they return. By putting labels, identifying and marking places or giving names, which is necessarily part of any culture [...]". (CLAVAL, 2014, p.19).

> [...] there are moments when the sun disappears in the bush, the sun disappears but the wind remains, then yes, just look where the leaves shake, then you will know where to go; whoever is from my time knows what wind is blowing at that moment; but when there is no sun there I can only prepare my nose to pull the sea through the smell, then that is where I go [...]. (Verbal information)150.

When other geographical or landscape references are added for orientation beyond shadows, it is important to name them as a way to facilitate the identification and location of places. This can be done by taking into account names of geographical aspects and facts, faunal species and flora, which will serve as cardinal points for orientation. For example, go in the direction of the through the smell of water, move towards the wind to find the plain, a mountain, a street, junctions of trails among others.

For the residents of the city of Macapá, the process of reading and interpreting shadows is not always well known by everyone, that is, not everyone can read hours and get to the places from shadow orientation. Unlike the inhabitants of the village of Aqui, who dominate orientation references based on shade trees, as the statement follows:

> When someone gets lost inside the village, we use some common points and well known by the majority or place to situate, we always orient to some big and old trees, a river, a family cemetery, a coconut plantation or palm tree. I explain this way when someone wants to know how to get to a place, I give names of the trees, things, objects so that the person doesn't get lost and arrive well [...]. (verbal information)151.

> [...] the shadows and the trees wouldn't help me at all I would get lost, besides in a few avenues in the city there are names, when I want to orient myself and get somewhere I use the search on the mobile phone of the place and then the application takes me where I want to go [...]. (Verbal information)152.

The testimonies show that the two places live in different technological times using guidelines, time and geographical locations, depending on their cultural moment. The inhabitants of the village of Aqui, do not depend only on objects for reading and interpreting shadows, because when there are no landscape references, the shadow of the individual himself serves as a reference for orientation, as long as he has mastery of the geographical coordinate of the place

---

149 Subject H.F.C., 56 years old, ordinary citizen, city of Macapá, Amapá, Brazil.
150 Subject F.L.S., 69 years old, village of Aqui, in Massinga, Inhambane, Mozambique.
151 Subject Q.S.R. 69 years old, village of Aqui, in Massinga, Inhambane, Mozambique.
152 Subject C.S.L., 28 years old, ordinary citizen, city of Macapá, Amapá, Brazil.

which is obtained through routine and empirical observations of the references of the mobility of shadows and the sun. At sunrise, the shadow points west/west (O/O) because the sun is at east/west (E/O), during the solar midday (apino), it points at the lower part of the day making it shorter, it is the shadow turned on and at the end of the day it points to E/O because the sun is at (O/O) (CASATI, 2001, p.123).

The projection of shadows obeys the position of the Sun in the celestial vault, and at dawn and dusk, the Sun is in a position of the horizon that allows to see the Sun at the lowest part of the celestial sphere, causing sunlight to reach the objects at a low angle, causing long and extensive shadows. Due to its position of sun incidence at dawn the shadows will be projected westwards and backwards at the end of the day. At local noon, the sun is at its highest point in the sky (apine), creating an incidence angle of ninety degrees, that is, the solar incidence is made in a straight line from the vaulted celestial to the earth's surface, making the shade short, overlapping the objects that project it. It is the change of position of the Sun in the sky that changes the size and position of the shadow during the day.

In the domains of shadow and sun orientation, it is important to add toponymy, which consists in cataloguing or fixing names of trees or objects for walking, as has been the practice that "[...] nomadic societies, when moving within a circumscribed area, use the shadows as points for rest, camping after a long walk" (TUAN, 2013, p. 221), in this case, the shadows demarcate the cardinal points of the path orientating in which direction they should follow, because in order to know the direction one should walk in,

> [...] I need a communication about what was seen, naming the land or what was seen with a name. The creation of a grid of toponyms makes it possible to speak of places even when we are far from them, socialising the experience of the terrain and extending the sphere of displacement and exchange beyond what has already been travelled by the individual or his or her neighbours. (CLAVAL, 2011, p. 31).

> The shadow is one of the '[...] location grids of traditional societies [...]'. (CLAVAL, 2011, p. 59).

In addition to toponymy, social experiences make it possible to associate the rose of the winds with the senses, recognizing what is "[...] in front through sounds, rumours and smells what is behind, on the side, above the line of the eyes and we complete through touch what is missing [...]. (CLAVAL, 2014, p.15).

Orientation becomes indispensable when walking to distant and unknown places, requiring an abstract process based on references or points that allow one to define walking positions, in this case, the apparent movement of the sun in the celestial sphere and the behavior of the shadows of objects are observable everyday phenomena that can facilitate orientation. To do this, it is important to qualify points by naming them allowing for the identification of places, since "[...] to name places is to impregnate them with culture and power, to signal, to mark [...]"

(CLAVAL, 2014, p.204), from references which can be names of lands, towns and micro accidents of relief, by means of names of towns and cities which can help in orientation from shadows.

The demarcation of time through shadow requires knowledge related to the movements of celestial bodies (stars, sun, and stars) that help to control the change of hours, days, months, and years. (TUAN, 2013, p. 219), acquiring a function to regulate social experiences, functioning as the sundial and soul[153] that marks the time and dynamics of life. In this context, the shadow plays the role of a physical element or environmental variable of multiple meanings and simultaneously used to measure time by means of the height and length of its projection. As a resident of Aqui, time was marked by shadows:

> [...] my wife and I, when we were dating, would arrange to meet when our shadows are halfway or short. We knew very well that this moment would not get in our way because everyone at that time only cares about resting [...]. (Verbal information)[154].

Regardless of the height of the individual, thing or object, shadows reach half the height or length of the object or thing that projected it at the same instant, as long as these objects and things are located in the same geographical coordinate. The experience of the inhabitants of Aqui, for reading and interpreting the shadow clock, resembles Machado's (2013) [155]analema-like solar quadrant, in which the author presents an instrument composed of an elliptical scale, on which are marks corresponding to the times and a linear scale of dates, with an indication of the months of the year (PHOTO 26).

---

[153] In Platonic philosophy as Aristotelian the soul constitutes the shadow of the individual, the only one capable of understanding the numbering of movements because it alone has the capacity of reasoning, therefore the shadow/soul is the reason.

[154] Subject A.H.T. 67 years old, village of Aquiem Massinga, Inhambane, Mozambique.

[155] Sundial that allows to obtain the time from the shadow projected by a person.

Source: Adapted from Machado (2013).

The analytic solar clock is installed at the Casimiro Montenegro Filho Astronomical Pole (Itaipu/PR), presenting characteristics that are perfectly suited to the model of perception and interpretation of the apparent solar mobility and shadows of residents of the village of Aqui.

In the analytic when the user positions himself on the date scale, in the position corresponding to the day in which the observation is being made, his own shadow intercepts the time scale making it possible to know the time. The small novelty of the residents is that its instrument allows the reading of three positions of shadows and sun throughout the year, establishing moments corresponding to the solstices and equinoxes including the seasons.

Interpreting the explanation of the inhabitants of Aqui, we can say that, besides registering time, the analemático, presents months grouped in three bands forming two rectangular triangles, being that the first forty-five (45°) degrees correspond to the moment when the shadows and the sun are on the left, coinciding with the winter solstice, already the other degrees, occurs when the shadows point to the extreme right point, in the same instant that happens the December solstice and at the moment when they point the division of the triangles is the moment of the equinoxes. The instrument also makes it possible to determine the solar angles, approximation of local latitude and longitude, the date and time of day.

According to Machado (2013) for the use and exploitation of shadows it is important to respect the dimension and approximate geometric shape of projected objects or things, including the position in which they are in relation to the space that is intended to be used, and may have a

113

spherical, cylindrical, conical, ellipsoid shape, among others. Basic area per animal, if it is the case of oxen breeding, in the shade has been the object of controversy among researchers, as: 5.6 ^m2 for Bond *et al* (1958); 4.2 ^m2Buffingtonet *al (*1983) and 1.8 to 2.5 ^m2 for Hahn (1985).

Machado (2013), Jackson (2004), Feteris and Hutton (2000), Silva (2004), reinforce for the obligatory domains of the amount of solar radiation, sunshine, the position that the sun and shadows are in one place, in order to facilitate their use and exploitation in an easier way. In relation to the sun it is necessary to know the angle of elevation of the sun or the zenith angle (ψ), which depends on the position of the sun in the celestial vault, that is, the angle of solar elevation (θ) and the azimuth angle (α) of the sun in a certain place (latitude), the time of the year and the time of day. That is, the solar irradiance varies according to the angle of incidence of the sun's rays which forms the zenith angle from the interception between the local zenith and the sun's rays, varying in the inverse ratio with the zenith, that is, the higher the Z, the smaller the solar irradiance area and vice versa (FIGURE 8).

Figure 8 - Variation of solar irradiance considering equinoxes and solstices

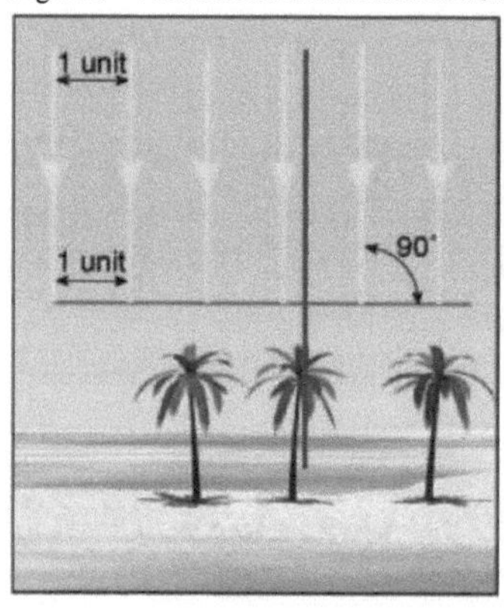

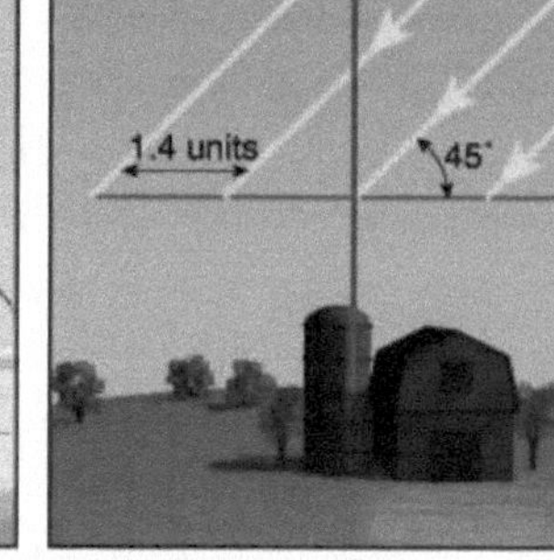

Macapá Z = 0°to ϕ= 0° | **Here Z= 45 to ± 45°**

Source: Organized by the author (2017).

The figure shows the positions that correspond to latitude: from Macapá at latitude zero degrees under the line of the equator with a zero declination, from the settlement Here latitude twenty-three degrees and declination of forty-five degrees. The red line indicates the zenith of the settlement, at noon Z can be obtained from the formula: $Z = - \delta\phi$. The figure represents equinoxes (Macapá) and solstice (Aqui) on 22 December, when the Sun is at its maximum declination to the south (summer solstice), it can be noted that at noon the Sun is in a perpendicular position over the central point, which represents the location of the observer, receiving the maximum intensity of radiation, see Photo 27.

Photo 27 - Shadow overlay in Macapá and Haqui

Source: The author (2015).

The first photo from left to right was taken in the village of Aqui, during the summer solstice, the last two were already taken in the city of Macapá at the drought or spring equinox, resembling the overlapping that happens due to the position in which the sun is in relation to the zenith, without giving the possibility of the objects projecting shadows out of themselves. This happens when the planet reaches a position in its orbit where the sun seems to be situated exactly on the circle of the equator or celestial tropic. The shadows of objects and things overlap in a null projection that symbolizes bound or zero shadow.

The zero shadow symbolizes the moment when the sun projects more radiation on the earth's surface, increasing in intensity the lower the latitude and vice versa at higher latitudes. In the Tropic of Capricorn the intensity continues to be greater because it is where the sun performs its maximum declination to return to the Cancer.

Another important element to evaluate the use and use of shadows is the photoperiod or sunshine (N) which corresponds to the length of the day, from sunrise to sunset, not including twilight (TABLE 5). Sunshine time regulates the behaviour of animals and plants, and it is important to study it because it facilitates the design of projects for the use and enjoyment of resources.

Table 5 - Solar Brightness in Macapá and Haqui

Latitude (ϕ) Place	Jan	Feb	Sea	Apr	May	Jun	Jul	Aug	Sep	Oct	Nov	Ten
00°02'18,84" Macapá	12,1	12,1	12,1	12,1	12,1	12,1	12,1	12,1	12,1	12,1	12,1	12,1
23°26'22" Here	13,4	12,8	12,2	11,6	11,1	10,8	10,9	11,3	12	12,6	13,2	13,5

Source: The author (2017)

The picture was produced based on the reading of the solar diagram, built from the *software Sol-Ar 6.2*, in which are the values for the fifth day of each month of the year, which allowed us to deduce that in Macapá the solar day is constant throughout the year, while in the village of Aqui the days exceed twelve hours of solar brightness between the months of September to March, recording values of no less than ten point eight (10).8) in the months between April and August of each year, being the longest in summer than in winter registering the shortest days in the month of May and the longest are registered on the fifth day of December.

According to Chang (1974), sunshine is also known as sunshine and varies in direct relation to latitude, that is, the lower the latitude the higher the value, also varying with the seasons, see Map 5.

Map 5 - Distribution of sunstroke in Macapá and Haqui

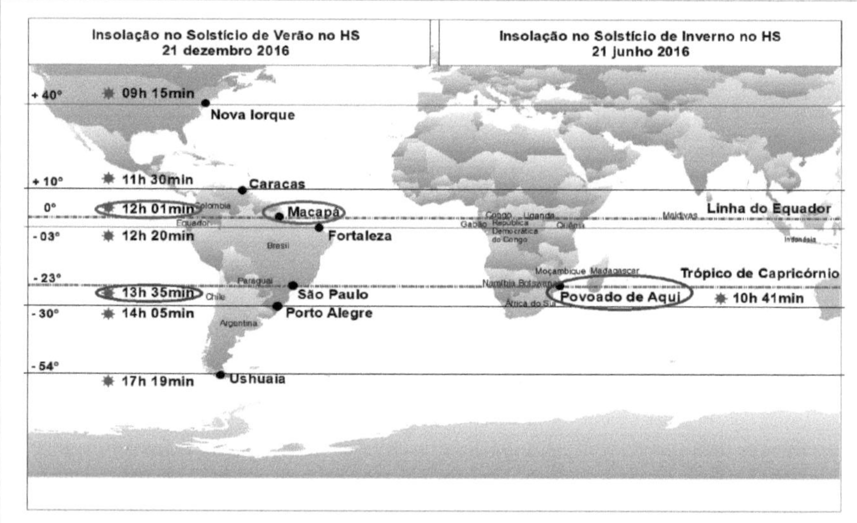

Source: The author (2018).
Note: Using HQIS 2.18.15 *las palmas*.

In all places situated on the Equator the duration of day and night is the same throughout the year, already in places situated on the Capricorn line, as is the case here, the duration is variable, and at the time of the solstices the day has a duration of 13h35min, meaning that the night will have only 11h25min.

The following highlighted testimonies describe the reality of Here about photoperiodism:

> [...] you know, these things usually coincide when the shadow is on the right side of my hut door, the swallows disappear, but it is good because they return to the village in cold weather, that is when they are important for healing ear pain from the cold [...]. [...] in December there is only mafura and cashew,

lettuce and other vegetables are well produced in cold weather because they
cannot stand cold [...]. (Verbal information)[156].

[...] hunger is intense when the snakes start to change their skin, the tambeiras
fall the leaves, their fruits dry up and burst, then the rats have bugs, stay in the
hole many days without even going out at night, [...]. (Verbal information)[157].

The length of the solar day, associated with the seasons of the year, causes behavioural
changes in the animals that immigrate, change colour, hibernate and in the plants start flowering,
leaf fall, drought, death, fruit production, reproduction, etc.

---

[156] Subject J.P.R., 52 years old, village of Aquiem Massinga, Inhambane, Mozambique.
[157] Subject A.H.T. 67 years old, village of Aquiem Massinga, Inhambane, Mozambique.

CHAPTER V

RESIGNIFYING EXPERIENCES

Shadows symbolize forms or objects in which values, attitudes, beliefs and social experiences are impregnated in the culturally lived space, that is, from the lived space the study of the perception of shadows is inserted in the cultural geography, which according to Claval (2007) is a set of instruments, artifacts or know-how and knowledge from which they mediate their relations with this medium.

Know-how and knowledge are the cultural tools that allow the individual to direct his senses[158] to clearly select and register phenomena that are of interest to external stimuli, selecting what is of value to him, blocking or leaving in the shade what does not provide satisfaction (TUAN, 1980, p.4).

Environmental satisfaction depends on the signs and symbols learned by the individual, when they become collective they constitute values and attitudes that forge identity, becoming culture of that group. Shadows are socially presented as signs to the extent that their reading and interpretation symbolize experiences or identity of a social group. Visualized from values and attitudes[159]that reflect culture, allowing strategies to be developed to explore the environmental system.

Attitudes and values are strategies that guide the social group to read and interpret shadows "[...] memorizing concrete images, visual impressions (landscapes), auditory impressions (noises), olfactory impressions (smells) [...]" (CLAVAL, 2007, p.194), which help the sensory relationship with objects and things that project them.

Shadow projecting objects and things constitute the environmental system of the place, bringing together fauna and flora where resources of recollection, hunting, agriculture, pasture, etc. are exploited, demanding attitudes that allow hiding places or animal tracks where the hunter can set traps while in the city these objects and things serve as indicators that allow the perception of urban thermal comfort to be analyzed.

Thermal perception is an environmental attitude driven by the psychic, which directs energies towards goals, making important in studies of this kind, the understanding of how culture articulates in relation to shadows projected by objects and things for the meaning of life.

Each society perceives and interprets solar and shadow mobility in its own way, as it happens in Macapá that are read to explain the solar obstruction in flats, public squares, parking garages, public roads in addition to being used to establish seasons or seasons. While in the village of Haqui, they are useful for identifying places for fishing, hunting, determining distances and geographical routes among others.

---

[158] The senses are sight, hearing, touch, smell and taste, known since Aristotle (TUAN, 1980, p. 7).
[159] It is a cultural position that is taken before the world, formed by a succession of long perceptions and experiences, therefore, the vision of the world results from the experience we have (TUAN, 1980, p. 4).

In the rural area, as is the case of the *Bastwa*, the orientation of housing obeys the eastern position, which not only serves to enjoy thermal comfort, but also to symbolize spiritual strength, goodness, joy and health. Using the front of their homes to recite a blessing and give "*moon*"[160] remedies to the newborns who take with their heads the sunrise. That is to say, when it is to take moon medicine, the child is put with his head facing east, never west, because the west symbolizes sadness or position that the dead take when they go to bury, also representing the end of the day, the moment when the sun disappears. The west also represents the final hour of the day's work, starting the daily rest for the living, a position that symbolizes eternal rest or physical death.

The night represents the greatest shadow of all, which corresponds to the moment of life when the dead come into action, acting like angels protecting or solving the convictions and aspirations of their relatives still alive. It is at this moment that the evil dead help the sorcerers by tormenting or throwing plagues upon the desires of the living, sometimes representing themselves in the form of dream-heavy, witchcraft, ghosts among other evil forms. But also, the greatest shadow is the moment reserved by the living to dialogue or receive guidance from their ancestors in the form of dreams, and receive guidance about the destinies of life. In this perspective the shadow is part of the night, it is in this moment that in cities like Macapá the squares are well used, because on the banks there is no more sun, there are lovers and families sitting in the shadow of the night, when the sun has gone to bed there is the dark that is the shadow.

The night orientations need to be aligned with knowledge that brings together the position of the shadows with the appropriate periods for carrying out a certain survival activity. In the case of agriculture, which requires strict knowledge about periods in the rainy seasons, it is necessary, domain of reading and interpretation of the agricultural calendar combining knowledge, to predict the month and days to happen in the rain.

In Macapá the shadows also symbolize rainy and dry seasons, with the March water equinox marking the flood season, raining torrentially during the March equinoxes. The Macapá people have used this experience to plan activities related to the rainy season. In this town, they use the summer solstice, and there is a detailed reading of the mobility of the shadows, to avoid agricultural neglect, because the farmer who cleans before the rains, runs the risk of being surprised by the rain with the land full of grass and without conditions for planting, while the one who clears after the rainfall happens, runs the risk of losing the crop and starving every year.

---

[160] Also known as clay or snail medicine, which must be administered twice a day, the first time at sunrise around six (6) hours and the second time at sunset around eighteen (18) hours. When applying medicine to the baby, the mother should kneel/collars position herself at the entrance of the residence, turning in the oriental position (if it is in the morning) or western (if it is in the afternoon) and apply the dosage. According to this tradition, a person who does not take this medication in early childhood is at serious risk of becoming epileptic, demented among other consequences that can affect his psychomotor development.

Among several utilities, the role of refrigeration and ventilation stands out, and shady places are used to install drinking water pots, constituting cafeterias for breakfast, known as mata-bicho in Mozambique, besides being places to eat fruit, lunch and leisure, as the next social subject says:

> [...] there are still families around the city who have the practice of cooling water and drinks using shades. Usually when they are drinks, they are buried in the wet underground or with wet sand under the shade of some tree. This experience is common in humble families, who on hot days want to drink something cold but cannot get a refrigerator, this happens here, I have witnessed [...]. (Verbal information)[161].

Knowledge about shadows also serves to determine distances, according to the speech of some social subjects, below:

> [...] measuring distances from shadows I have never heard of, but it is common to choose streets with many shadows to use during walks in the city. If an avenue has no shadows I don't use in the hot time I look for the one with the most trees, even if it takes the longest to get me to my destination, so the shadows make up for the distance [...]. (Verbal information)[162].
> [...] since childhood we have learned to fix in the head the amount of shadows and the species of trees that give shadows along the way, so it is easy to know how many shade trees there are from one place to another. This helps a person to have the slightest idea of the distance he is going to travel and the possibility of burning in the sun or not until he reaches his destination [...]. (Verbal information)[163].

The following table (6) provides estimated data offered from the Haqui village headquarters for some points, which are more frequented by residents. The schools indicated are the points used for the exercise of formal education, but also the places where residents hold meetings to participate in vaccination campaigns and advertisements on the promotion of children's life rights and gender equality.

Table 6 - Distance measured from shadows

Place of departure	Destination	Number of shadows	Distance in km
Povoado de Aqui Headquarters	Guizugo	11	17
Povoado de Aqui Headquarters	EPC Mahocha	23	8
Povoado de Aqui Headquarters	EPC Malova	32	9

Source: The author (2016).

Most residents do not use either public or private transport to travel to the various places that make up the day. The walks are made on foot following trails called paths, which require shadows along the way to go resting, protecting themselves from sun heat, it is in this perspective

---

[161]Subject S.M.M., 38 years old, SETUR, city of Macapá, Amapá, Brazil.
[162] Subject G.M., 48 years old, ordinary citizen, city of Macapá, Amapá, Brazil.
[163] Subject V.J.M. , 54 years old, village of Haqui in Massinga, Inhambane, Mozambique.

that you fix along the itinerary shadow points that determine the distance to be covered on a path. For example, the Mahocha Primary School (EPC), located in the north of the village of Aqui, until arriving at the EPC from Malova to the south, is about fourteen (14km) kilometres away along the EN1, with about one hundred and fifty trees with good shadows of cashew and mango trees. What happens is that the majority of the inhabitants of the village of Aqui, do not like to walk along the trail on the side of the road, because they drive at speed, so they teach the children, to use roads or trails inside, although there are only coconut trees that do not project good shadows, but the way is to walk there, it is worth catching the sun, have headaches, than losing a child by car accident.

It is noticeable that the villagers walk and demarcate distances based on shadows when describing the amount of shadows that can be enjoyed by walking along National Road number one (1) to the desired village and lamenting the lack of clear shadows on the hiking trails inside the village, far from the Road.

In the city of Macapá there are people specialized in architecture and interior design to improve the thermal comfort of residences, to avoid socioeconomic disturbances to the residents, as stated by the social subject:

> [...] some think that just the air conditioning is enough to solve the heat of Macapá, but this can be expensive for the pocket, because more than one air conditioning may be needed in one house. The solution is to know the orientation of the house in relation to the sun and the shadows in relation to the windows and rooms of the house [...]. (Verbal information)[164].

> [...] there are people specialized in building huts and shacks, known as *muyaquewatihilho,* that is, a connoisseur of building techniques involving the aesthetics of projecting shadows on the outside and good lighting on the inside through low shading on the inside of the hut. Another requirement is that the builder has mastery of the geometry of spatial occupation, that is, knowing how to occupy the space in the yard so that the shadows are not projected out of reach of the inhabitants of the house [...]. (Vernal information)[165].

The house should be facing the apparent sunrise, with the leafy trees planted two metres in front, which help in the ventilation of the inside and outside of the house through the projection of shadows, these practical skills, usually help choices of shadow room for classes, meeting rooms, shady rooms for parties, spaces for traditional rituals, places for dating among other utilities.

> [...] there is a practice of planting *xitsalala*[166], it presents a partial or less dense, ritualized shadow, as of brainwashing, that is, the exposure over its shadow has a therapeutic power helping to forget evil things and bring back in the memory

---

[164] Subject L.B.A., 47 years old, UNIFAP, city of Macapá, .
[165] Subject B.C., age not identified, technician of Massinga's Economic activities.
[166]Scientific name of the Strychnosspinosa tree.

good memories. Yellow acacia is little disseminated despite projecting good shadow, spreading the evil myth that the most important person in the family - the man said to be the father of the family, multiplier of the species, who is never sterile - can die. Discouraging planting in the middle of the yard, the reason for the prohibition being that the thick and long roots of the yellow acacia can create cracks in the foundation and wall of the house and collapse [...]. (verbal information)[167].

Knowledge about shadows includes knowledge that is applied to fishing activity as the residents speak:

> [...] one thing that many fishermen know around here, is that leaving the fish under the moonlight will get ground up, so you have to put it over a dark place or over the shade. Now fishing thinking about the shade I don't think they know anything because I've never heard of anything related to fishing and shadows, I've been working with the fishermen of Amapá for some time [...]. (Verbal information)[168].

> [...]the techniques of capturing animals in hunting include practices related to knowledge of the places where they often shelter for thermal comfort. Animals during the warm season use the shadows to lie down when they are in a comfortable situation while under discomfort they are normally standing and it is difficult for hunters to catch them. Knowing the path of the shadows are also used for structuring annual cycles [...]. (Verbal information)[169].

(LATOUR, 2012, p. 52), which presents form, functions, senses and interpretations resulting from cultural perceptions, becoming "[...] real-abstract and real-concrete totality through social forms [...]" (SANTOS, 1996, p. 98), allowing functionalities or a system of meanings, as the following statement states:

> [...] a person's shadow represents their soul, it is in close connection with ÒkúÒrun and *Àgbagbá*, it is in the shadow where the essence of life is, the body is only a container that surrounds the organs and not the soul [...]. (Verbal information)[170].

> [...] for us shadows represent the soul of our ancestors, with a healer we sometimes invoke and talk to them, but in truth it is not with their shadow that we incarnate in the healer and represent ourselves through him to give guidance on how we should proceed while we live. When the ancestor appears sometimes he wants to dance with us, sing, cry, at last he behaves as he was when he lived [...]. (Verbal information)[171].

To share different forms is to assume that the interpretation of them is a social construction that can bring people closer together or detach them, highlighting those that represent the bad things in their lives as haunting, developing feelings of topophobia, fear or desire for

---

[167] Subject A.L.C., 54 years old, village of Haqui in Massinga, Inhambane, Mozambique.
[168] Subject L.M.N., 47 years old, ordinary citizen, city of Macapá, Amapá, Brazil.
[169] Subject B.C., age not identified, technician of Massinga's Economic activities.
[170] Subject A.B.C., 35 years old, ordinary citizen, city of Macapá, Amapá, Brazil.
[171] Social subject T.M.M. , 41 years old, village of Haqui in Massinga, Inhambane, Mozambique.

detachment, being considered by psychicists as symbolizing heterotopias[172] (FOUCAULT, 2002 apud CLAVAL, 2014, p.60).

> [...] in our language shadows sometimes mean bad things, those that nobody likes, bad memories, but also thieves around the city attack residents when they are in the shadows of public squares. Therefore, it is not common to be alone or insolate in a shadow, at risk of being mugged, these are some bad things of the shadows around the city [...]. (Verbal information)[173].

> [...] the dark creates fear, especially when it passes in a cemetery at the risk of meeting a ghost. But also because it represents the time when thieves circulate freely in the city [...]. (Verbal information)[174]

> [...] in the dark it is difficult to notice where there are scorpions, snakes and other animals that can take advantage of the dark to attack man. But it is also frightening because we still have in our heads, that armed bandits usually circulated here at night in wartime, this still prevails in our heads [...]. (verbal information)[175]

In this context, Turnbull (1965) and Tuan (2005) state that fear of darkness or shadows is a human creation, and there are societal groups such as the *mbuti* pygmies of northeast Congo that their concepts of fear, threat, suspicion and anxiety are greatly diminished, and only positive aspects of the shadow emerge for them. Living in a broad canopy forest that hinders the intense sunlight, that forms in its intense shade interior establishing thermal comfort, abundance of mushrooms, nuts, pods, roots and fruits. For others the shadows are habitat or burrows of game animals, such as: monkeys, bush pig and elephants serving as fast hunting sites.

The fact that the shadows serve as places of concentration of animals for hunting allows them to serve as probes or locators of places of fast hunting leaving time throughout the day for other tasks, such as: having time to talk, set traps, play with children, sing and dance. As described by J. M. L. of the village of Aqui:

> [...] rabbits, they like leafy bushes and don't like shadows of big trees. Partridges and rats hide in the shadows of elephant grass [...]. (Verbal information)[176].

> [...] when we were kids we knew very well that the best places to shoot birds were on broad-top trees with fruit, such as hoses and juggernauts, that's where even without good aim we could only shoot in the direction of the flock of birds that gathered to try to eat fruit. Generally the broad-top trees have good shade, there the birds enjoy food and shade from the tree in hot weather [...]. (Verbal information)[177].

---

[172] Areas of difficult survival, where contagious patients, insane people, delinquents or quarantined people are isolated.

[173] L.B.A., 47 years old, UNIFAP, city of Macapá, Amapá, Brazil.

[174] Subject A.C.D. 27 years old, city of Macapá, Amapá, Brazil.

[175] Subject S.S.D., 36 years old, city of Macapá, Amapá, Brazil.

[176] Subject J.M.L., 39 years old, village of Haqui in Massinga, Inhambane, Mozambique.

[177] Subject S.D.A., 57 years old, ordinary citizen, city of Macapá, Amapá, Brazil.

Shadows create in the mind two worlds, one consisting of places of everyday life, that of the eminence constituted by the shadow as a concrete and abstract physical phenomenon, and the other of the imagination which is transcendental formed by the world of ideas, values and attitudes related to the invisible.

Natural phenomena such as: the vegetative cycle of plants, cyclical change of seasons, alteration of the appearance of the sky creating darkness, temperature variation can be interpreted from the shade as a visible concrete physical phenomenon. Shadow as something abstract represents the soul, the divine power, the unconscious of Gestalt and transcendental psychology which constitutes structures of social organization, as some residents tell us:

> [...] I know that in the interior of some municipalities of Amapá, people still respect the talk of women and men, reason to establish where to stay each group, in the city of Macapá there is no such practice. Even because tying a hammock can be done separately, a place for the man or for the grandfather, but most of the city what it really likes is air-conditioning [...]. (Verbal information)[178].

> [...] children always have somewhere to play away from big people, but women never stay in the same shadow with men. The spirits are invoked in a specific shadow. When a young girl gets pregnant out of wedlock she also has a place to solve a problem [...]. (Verbal information).

This is how the shadow groups its members according to age, family prestige, sexuality, placing individuals according to "[...] male or female gender [...]" (HENRIQUES 2011, p.73). In this perspective, the shadows socially group the members of a family according to sex (shadow of men and women), age groups (shadow of children, adults and elders), shadow of the water where the pot is fixed or of the "refrigerator", shadow of the dead, of the healer, among others, requiring knowledge about their trajectory throughout the day.

For Head (1995), Costa *etal* (1997), Carvalho *et al* (2002), Primavesi (2006), Dias-Filho (2012), in these societies the perception of shadow plays a very important role in the promotion and sale of certain species of plants and houses.

> [...] it is important to know the position of the shadow in the rooms and other places of the house, this values very much the price of buying and selling flats. The sale of flats according to the shade also adds to the question of the type of forestation that adorns the house, if the plants are native or imported, if the plants offer good landscape and good shade to the place, this counts a lot when selling or buying a flat [...]. (Verbal information)[179].
> [...] what we should plant should please by giving food through fruit, but it should also have good shade to use when we are walking around or when we want to rest even in the field we must always have a good shade tree to sit on.

---

[178]Subject E.L., 53 years old, ordinary citizen, city of Macapá, Amapá, Brazil.
[179]Subject H.T.C., age not identified, urbanization, city of Macapá, Amapá, Brazil.

<blockquote>Therefore, the choice of plant species should obey the question of producing fruit and shade at the same time [...]. (Verbal information)[180].</blockquote>

In the semi-arid like in the interior of the Massinga district, with infertile soils the production of vegetables and other cultivars is done in shade. Even in fertile soils such as the town of Macapá and the village of Aqui, photo-period and photo-tropism domains are necessary,181which influence the growth of each plant (height) and the position of the shade throughout the day and the year. We show that

<blockquote>[...] the pruning of plants by the city must be done when it is autumn, when the leaves fall, they cannot be cut during the flowering season under the risk of losing the fruit harvest and the aesthetic given by the flowers to the city [...]. (Verbal information)[182].

[...] each plant has its own time to be planted, for example, cashew, coconut and other plants are usually planted from December to February. Outside that time it happens that they grow slowly, their production in the future appears to be impaired by heat or cold. The plants also show different joy in the moment of heat and cold, because of the sun which is little [...]. (Verbal information)[183].

[...] Plant cultivation is not only the manifestation of aesthetic sense, nor only a complementary economic strategy, but also a source of meanings and practices in which *status*, conflicts and aspirations are constantly negotiated and manipulated among the genera. They end up functioning as efficient tools of negotiation within and between domestic units in which economic survival, gender difference, social *status* and emotions play fundamental roles. (MURIETA; WINKLER PRINS, 2006, p. 290).</blockquote>

As a source of meaning, practice and status for some public institutions, they represent power hierarchies based on thermal comfort, having as an example the shadow of the commander in the Matalane Police Practical School - Mozambique, where only the commander and other hierarchical superiors hold the prestige of landing on the leafy Tiziva tree[184](PHOTO 28).

---

[180] Subject H.I.J. 63 years old, village of Haqui in Massinga, Inhambane, Mozambique.
[181] Competition from the shade of sunlight between plants.
[182]Subject H.T.C., age not identified, urbanization, city of Macapá, Amapá, Brazil.
[183]Subject V.J.M., 54 years old, village of Haqui in Massinga, Inhambane, Mozambique.
[184] In Xitswa, it refers to the name of the tree or plant known scientifically as *Dialiumschlecheri*.

Photo 28 - Shadow of officers at Matalane Practical School (MZ)

Photo: The author (2017).

The photo was taken during the shade or summer solstice in December 2016, about half a local day while the shade overlaps the tree. Another example is the choice of an accent or chair on public transport, obeying a position that allows during the trip to avoid the incidence of sunlight, using the shade as an element of choice/attachment or not to the accents inside the bus, adding in the shade the category of "[...] topophilia or topophobia [...]" (TUAN, 1980, p. 7), symbolizing "[...] place of invisible beings [...]" (HENRIQUES, 2011, p.103). Identifying haunted environments as geographical spaces of evil, characterised by the presence of ghosts, curses, etc.

> [...] I don't know much about the negative aspects of shadows, even though I am a university student, but I can kick. But I don't know very well why one compares bad things to shadows, maybe it's because it's a calm, quiet, cold place, that's a reason for competition from these places. The competition creates disputes, because everyone prefers to succeed, that is dangerous. Another thing is that the shadows of our city of Macapá are places that can only be enjoyed in groups, alone it creates fear because of thieves. The shadows represent the dark, place of the night, haunting, distrust and betrayal, so negative things are compared to shadows [...]. (Verbal information)[185].

> [...] as my colleague said, the deceased are sometimes bad when they die angry with someone and can cause ecological crises and serial bad luck about their life, so it is always good to have a party once a year, usually the ritual is done under a tree shadow [...]. (Verbal information)[186].

The belief in miracles causes most Ba tswa to perform peace rituals, good harvests on the leafy shadows, usually from trees considered shelters of ancestors' spirits, with powers to protect

---

[185]Subject X.B.E., 24 years old, ordinary citizen, from the city of Macapá, Amapá, Brazil.
[186] Subject A.H.T. 67 years old, village of Aqui in Massinga, Inhambane, Mozambique.

the living from calamities, pouring blessings and successes on all they desire. Symbolising topophilia (TUAN, 1980), which manifests itself in the extent to which shadows are regarded as good places for the life of the deceased, as well as places of abundance of animals for hunting, honey extraction, sites for recollection or harvesting of fruit including from edible leaves, among other benefits taken from shadows.

For the use and use of shadows it is useful to know the height of objects and things that project shadows, the position in which they should be allocated and the material that constitutes the object that provides shadows. But it happens that no one has paid attention to the size of the trees placed in the avenues of Macapá, which despite being leafy, their shadows occupy the whole road, forcing pedestrians to walk in the middle of the street to take advantage of shadows, causing accidents. Another thing, is that in the city, when it is mango time it has fruit beating the cars and people on the road or it has dirt in the streets, sometimes it is fruit trampled by cars, it is so absurd among shadows, cars and people in this city.

In the village of Aqui, the planting of trees in addition to species selection, obeys domestic ventilation, choosing the right place for shade. For example, the coconut tree is not recommended for the central part of the house because of its fruit, which can hurt when it falls and does not offer good shade. Even when the shade is from a hut or stall it is important to know how to build the part of the house that will give shade, respecting the height of the house, which is advised to be very high, because when it is short it does not help at all for good shade.

Findings of Macapá residents show that shadows seem to do the same thing every day of the year, being long in the morning and in the afternoon, shorter around noon. Even without changing the shadows what we all know is that in March it will rain a lot and in September it will be hot and very warm. Everything to do with rain already knows that it will happen in March without fear of making a mistake, the hot things like going to the beach, tanning and traveling by the river will be in September without fail, so on. This is also the case in the village of Aqui, where the residents are attentive throughout the year to identify the right moments to knock down the bush and burn as a way to clean the land, to know the position of the sun, the shadows and the type of clouds that are circulating in the sky. That is, although the shadows every year are long in the morning, short in the afternoon and long at the end of the day, the Aquians dominate the right times for rain. Everyone also knows that when the bright white clouds appear high up there, it won't be a week without rain. If the white clouds exceed ten (10) isolated groups, then the rain will be greater. And that in the following week the black clouds will soon appear well loaded with rain running so fast, so it's a matter of hours, it will rain. At that same moment, some elderly and well-confirmed people start complaining of body aches. Some donkeys bray desperately, the cocks cluck intensely, the birds flap their wings regularly, ants and swallows appear soon to announce that the rain has arrived, minutes later it's rain that doesn't end in the whole village. During this period, some tidy up the roof of the houses so as not to spill, others clean the land and start sowing, some clean up cisterns or rainwater reservoirs so as to be found by the rain well-cleaned and to conserve water, etc.

> [...] when the shadow is in the middle, *tsuvukagambo*[187] the sun rises more or less in the direction of the Capricorn landmarks, there begins the noise of the croaking of frogs in the marshes and ponds, it is a time of much malaria, everyone is trembling to knock their teeth, many die at this time, this time there are many diseases there and they reach the village, like cholera and malaria, but there are also many fish from the river and the sea. The cold starts, it is the time to eat beans, apples or fresh corn, peanuts, sesame, at this time the dogs do not sleep all night screaming because the males take advantage of the heat to make coitus. There are many *rats*[188] to hunt and eat. There are also beetles from the leaves of trees, which are edible *matamane*[189]. When the shade moves a little to the right, the drought begins, during this time there are many clandestine paths due to the trampling of people and the grass does not resist dying, it lives on some edible herbs and plants that appear on vacant land or in the bush, such as: *cacana, mboa,*[190]*tinembenemba*[191], *xindokomelana,* etc., in the bush there are some wild fruits that resist drought such as *massala, macuacua,* cacti, etc., it is time to fish too. Once again the tangerines, oranges and cashew appear when the sun shade is well towards the south. But what you should know is that the sun and the shadow are always together, the sun rises in the morning to the east and lies down to the west throughout life, the shadows also in the morning are long and projected to the west, at noon they are very close to us and the afternoon are long but projected to the east [...] (verbal information)[192].
>
> [...] a farmer who is distracted from sowing long before the rains, when the plant sprouts if it is cereal will take two weeks to flower and a farmer who sows two or more weeks while he is still cleaning the land after the rains, the corn or beans will have an early flowering and his production will be impaired. So I just have to see my male duck that black one, I know it will not fail in two days I start to prepare the field and the seeds to sow in time [...]. (verbal information)[193].

From this perspective, daily life here is marked by great cultural diversity, where the Ba tswas people of the village have developed their own techniques for relating to the environment, organising their organisational structures over natural cycles.

> [...] the rain has fallen a lot in the normal time when the sun begins to rise in the direction of chiduca, it is also the time of a lot of orange, a lot of food; but when it begins to rise more in the left direction, everything is already dry, there is only cacao, everyone eats what they produce in the machongo [...]. (Verbal information)[194]

---

[187] Centralized at the eastern point.

[188] Domestic mouse, known by the scientific name *Praomysnatalensis*.

[189] Leafworm.

[190] Known as *tseke* in the rogue language of Maputo and scientifically *Amaranthusspinosus*.

[191] Macao beans scientifically known as *mucunapruriens*.

[192] Subject J.V., 53 years old, village of Haqui in Massinga, Inhambane, Mozambique.

[193] Subject V.M.L., 46 years old, village of Haqui in Massinga, Inhambane, Mozambique.

[194] Subject L.G.A., 43 years old, village of Haqui in Massinga, Inhambane, Mozambique.

The months are named according to the events of the year, e.g. Damby is a[195]time of food abundance, and the torrential rains in the village do not cause flooding, due to the characteristics of the soils (sandy), altitude and continentality in a place with a weak hydrographic network.

The position and size of the shade at each moment is associated with the occurrence of ecological events and human action, for example, the December shade is associated with the period of ripening of fruits, harvesting and sale of cashew nuts, mafura, purchase of household utensils and other goods for family use and consumption, as it is associated with the abundance and festivities.

When the shadows overlap the objects that project them, they symbolize the inversion of the sun's direction of march, which begins the apparent journey northward. During this period, the main local fruit trees (cashew, mafureira and coconut) are at their peak, this happens during the nhanhany[196] or solstice of December, a time of fruit premises, coinciding with Christian Christmas and torrential rainfall that begins in the third week of December marking the end of the annual cycle of food production for survival.

The end of the annual cycle of food production for survival is scheduled by reading and interpreting shadows, determining moments of each activity, serving as an indicator to situate residents on where to hunt or fish, prediction of natural events, such as: approaching the season of strong winds, rain, drought and pests, as well as to control biological cycles relating to plants, fish and animals, including variations in the water balance prevailing for the agricultural calendar.

The agricultural calendar and other family survival activities such as artisanal fishing, hunting and family agriculture require a combined knowledge of the behaviour of thermal rainfall variations, the water balance resulting from excess water (more water that saturates the soil) and water deficit (less precipitation and lack of water in the soil). The village has a water deficit all year round, even during the rainy months, but never reaches surplus water or floods, and the shadows allow for the organization of production structures, including the establishment of natural cycles through shade to the ecological events of the village (TABLE 7).

Box 7 - Shadow movement over the year according to Haqui residents

SUN POSITION	MONTHS OF THE YEAR		ANNUAL CYCLES
	Xitwsa	PORTUGUESE	
This	Mussunguly	January	First winter harvest
This	Mahanzanye/Nyanyanyani	February	Second winter harvest
This	Nyanyakulo	March	First extractivism in the shadow cycle
This	*Zivamussoko*	April	Winter Seeds

---

[195] Flood or flood in *xitswa* language.
[196] December in *xistwa*.

West-Northeast	*Mugatchiye/Pwitawuxika*	May	Third winter harvest
West-Northeast	*Malataxikwinyane*	June	Winter rain
West-Northeast	*Mawuwane*	July	Drought of plants and animal husbandry
West-Northeast	*Mhahure*	August	Animal and plant extractivism; Handicraft
Lés-Sudeste	*Ndzati*	September	Male work
Lés-Sudeste	*Nhlangulo*	October	Regeneration of plant and animal species, *matoman* season
Lés-Sudeste	*Hukuri/Khumbo*	November	Seedshop
Lés-Sudeste	*N'wendamhala*	December	December harvests; Fruit Premises Parties

Source: The author (2017).

According to the inhabitants of Aqui, the shadow always takes a projection obeying the position in which the sun is found throughout the year, and in the first four (4) months of the year (January, February, March and April) the sun is projected in line with the East (E) cardinal point in the months of (May, June), July and August), is positioned on the left in relation to the first moment, we designate this stage as Lés-Noroeste (ENE), the third and last phase, happens in the months of (September, October, November and December), when the shadow is to the right of the point East, designated Lés-sueste (ESE). The three moments of the year coincide with equinoxes and solstices, see the comparative scheme (SCHEDULE 1).

Diagram 1 - Natural cycles with the solstices and equinoxes

Source: The author (2018).

The natural cycles are equivalent to a common or leap year, generally used in urban areas such as the city of Macapá. The first stage of the natural cycle through the shadows, has three sub-stages, the first of which starts with the cleaning of the *field*[197] by means of hoeing198 to save crops of corn, peanuts, beans and other cereals, which correspond to the first winter harvest.

Once this phase is over, the second sub-stage begins, consisting of vegetal and animal extractivism, harvesting fruits and wild leaves, such as: (*mandocomela/mahungwa*[199], *mambombo*[200], *mavilwa*[201], *Mahlala* ,[202]*Macanhu*[203], *Macuacua* ,[204]*Tinembe-Nembe*[205],

---

[197]Cultivated land, generally known as "roça" in some parts of Brazil.

[198]It is understood as a process of ploughing to remove weeds, which are born spontaneously in the *field/field* and can interfere negatively with agricultural production.

[199] *Apocynaceae* family.

[200] *Celastraceae* Family

[201] *Vangueira infanta*

[202]Strychnosspinosa

[203]Scleocaryacaffra

[204]Strychnosgerardi

[205]Casiadelagoensis

Tichindzo[206]), there is the opening of hiking trails through the woods, due to the constant trampling for food. The third sub-strip, corresponds to the fall of rains, sometimes happens in this period the *zivamussoko*[207] sowing of winter, usually is for beans, corn, yoke beans and vegetables. The third subetape, called *mugatchihé,*[208] coincides with the winter harvest. It also has plenty of fruit, and you can have lunch, a snack and dinner of juices and fruit, without having to make cooked or baked food.

The second natural cycle of shadows presents its subdivisions, namely: the first happens when the shadow is in (SE), it is *malatachikuinhane,* literally equivalent to sleeping hungry or a change burning all night to protect yourself from the severe cold. Some plants die of the cold and others just lose their leaves, some animals that can't stand the cold, to conserve their energy and save themselves from the lack of food, such as squirrels and rats go into hibernation.

The second subdivision of this cycle begins with vegetal and animal extractivism, developing craftsmanship, taking advantage of plants that are in the regeneration phase through the appearance of leaves and flowering, presenting qualities for spinning, allowing the manufacture of ropes of tambourine, sisal, palm trees, manufacture of baskets and wallets of straw. Because the trees are in a good time to be used, the process of reform of the haystacks begins, construction of barns, houses, manufacture of cloths called *mutchales*[209] and oils through coconut and peanut.

The third subdivision of the second cycle of shadows happens when the shadows are on the right (NE), starting with the process of exchange and preparation of land that was lying fallow, cleaning through controlled burning, sometimes accidents happen where fire invades other areas creating human and material damage. There is the exchange of reptile skin, snakes *tchemulate*, birds sing, the fruit of the tambeiras[210] dry and burst causing banging, it is the period of hunting birds and producing vegetables in the shadows of trees and *males* because in the sandy highlands are dry.

The third cycle of shadows is represented by the beginning of rains, sowing of cereals, cashew nuts, control of partridge folds and other trees in order not to remove the sown cereals. Other fruits such as pineapple (papaya), banana, minutes appear. The favorite dish is *kulula* or

---

[206] Scientific name *Phoenix redinata.*

[207] It is synonymous with *nkwuamulambo,* or cyclone, strong winds with torrential rains.

[208] It means food diversity in Xitwa, that is, a lot of food choice.

[209] A type of local factory cloth, usually used to carry children in early childhood, a plant that was used to make these cloths is in extinction and is still resisting one in life.

[210] Brachystegia family tree, typical of the miombo forest in Mozambique.

konolela, usually many inhabitants of Aqui complain about bile[211]at this time is that *xitunguti*[212]appears.

Outside the *xitunguti* that serves as a bioindicator, there are other forms of weather forecasting or interpreting atmospheric phenomena, such as: knowing how to read and interpret the circulation of air masses, precipitation and temperature variations, geographical orientation from the wind rose based on the position of shadows, applying daily experiences for the application of perpendicularism in determining directions and azimuth for geographical orientation through the wind rose based on shadows (FIGURE 9).

In the city of Macapá, the narrations did not allow the building of a rose of winds giving birth to shadows, but it is possible to notice that at some times of the year, some experiences coincide with equinoxes in March and September, as well as with what is called spring and autumn in the city.

Figure 9 - Haqui village wind rose

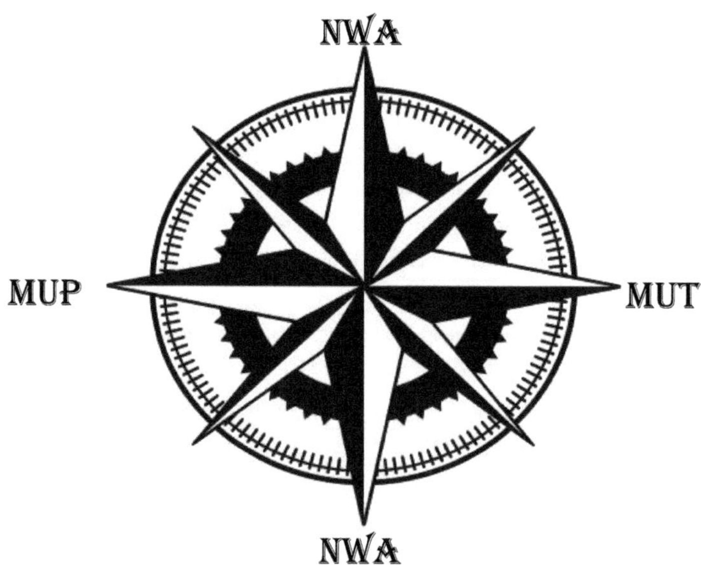

---

[211]Gallbladder disease is more common in women, overweight people, people with gastrointestinal problems and high blood cholesterol levels.

[212] Type of plague can be grasshoppers or insects that announce the approach of a famine period, that is to say they are bioindicators of a prolonged drought period.

LEGEND

Language Portuguese	Language Xitswa	
N	Nwal	Nwalungo
O	Mupg	Mupelagambo
S	Nwan	Nwandzonga
L	Mutg	Mutsuhukagambo

Source: The author (2018).

Nwal/Nwalungo corresponds to the North Cardinal point, *Mupg/Mupelagambo* the bridge or West, *Nwan/Nwandzonga* the South point, *Mutg/Mutsuhukagambo*, to the East or East point.

The cyclones affecting the region where the village of Aqui is located in the 21st century include Japhet, Favio and Dineo, which took place in 2003, 2006 and 2017, respectively, between the months of February and March, due to the low pressures of the Mozambique Hot Canal (CQM), which devastated the northern and southern coasts of the country. On this subject, the following are testimonials from the residents of Aqui:

> [...] shadows cannot be read and interpreted in an insolate way, it is necessary to add to them dynamics of other things that happen in our life, for example, the circulation of clouds and winds, causes harmful ecological events that no one can explain by shadows alone. For example, the winds called *moyawaMupelagambo*[213]that happen in the shade (CE), are rarely of occurrence, but when they appear they are dangerous because they are accompanied by *Nwuamulambo*[214], as it happened when we lost relatives, destroyed houses and many goods, these winds are very strong and with heavy rains, big hail falls, which if you hit someone's head, it kills right away. Other winds, called Nwalungo,[215] blow to *Nwadzonga*[216]announcing torrential rains, usually happen in the shadow phase (NE), and when they blow in the opposite direction, they mean, calm and dry, they happen at the moment when the shadows are leaving (CE) to (SE). The *Mutsuhukangambo*[217]wind is accompanied by showers[218]. (Verbal information)[219].

> [...] shadows also serve to evaluate the quality of resources, for example, the quality of raw material used in the production of baskets, nets, ornaments and hunting traps. When taking a palm tree to manufacture *Fridays, xipatchi* is recommended to be exploited in healthy trees over places of thermal comfort allowing to establish aesthetic or qualitative standard of objects from it. This

---

[213]Westerly winds in the Xitswa language.
[214]In Xitwsa language, it refers to strong winds accompanied by torrential rain with thunderstorms and lightning, it can be cyclones.
[215]North winds in xitwa.
[216]South winds.
[217]East Winds.
[218]Rainfall with rapid variations in intensity and sudden alternation of cloudiness.
[219]Subject V.M.L., 46 years old, village of Haqui in Massinga, Inhambane, Mozambique.

> means that when the rainfall rates are high, the trees become leafy creating
> thermal comfort from the shadows, constituting better animal shelters for
> hunting and fishing. The shoals are concentrated in shady places of leafy trees
> that during the transition from winter to summer lose their leaves, becoming
> better places for the production of *placton* that feeds the fish [...]. (Verbal
> information)[220].

Although Macapá does not use the shadows to schedule the year, it uses them as elements of tourism, from the contemplation of the overlapping of the shadows as an activity to promote tourism, which takes place in the monument Marco Zero, already in the village of Aqui is not yet part of its daily life, happening sporadic visits of some astronomy admirers and tourists to take pictures of the signs that symbolize the passage of the imaginary line of the Tropic of Capricorn through the village.

Experiences represent the understanding of how each geographic unit of study organizes itself and builds meaning from the shadow, which constitute its ways of life or its way of understanding and explaining the world. In the case of thermal variability, shadows are joined to the things that project them functioning as ventilators simultaneously as hydrological watering systems that, besides softening temperatures, retain the objects that protect them from condensation of dew and fogs that feed the plants giving salts and minerals, but are also organizers of everyday structures that aggregate all forms of life of these population groups.

---

[220] Subject G.M.C., 49 years old, village of Haqui in Massinga, Inhambane, Mozambique.

Because of their geographical location on the imaginary lines of Ecuador and the Tropic of Capricorn, the residents of Macapá and Haqui have developed their own experiences of the perception of equinoxes, solstices and shadows, as socio-environmental concepts of their day postpone.

From the shadows, the social subjects of the areas of study defined the shadows from concepts related to what is transcendental that which is beyond known or visible limits, often attributing to a divine or sacred mystic being symbolizing through the soul of the ancestors. This observation recalls the idea that the perception of the inhabitants of Aqui e Macapá is that shadows are immaterial, abstract, using them to construct concepts and theories about their daily lives. As has already been mentioned by the philosophy of Plato (428-348 BC) and St. Augustine (354-430), that all things and objects in the real world are shadows or ideas of eternal forms, still studied by various scholars today, such as Zweig and Abrams (2011) who refer to shadows as the centre of the personal unconscious or the soul of the individual, used to make psychological therapy in people or sources of formulation of scientific concepts and theories as Casati said (2001).

Another observation is that shadows are material or visible phenomena that serve as sources regulating ambient temperatures providing thermal comfort. The equinoxes and solstices represent moments in which the shadows overlap the objects marking changes in the seasons, in the intensity of the solar radiation, in the duration of the sunshine or sunstroke, allowing the inhabitants of Macapá and Haqui, to organize their activities in accordance with the concepts of photoperiod and photoperiodism, hibernation, circulation of air masses, preservation and conservation of species. This is to say that the residents have developed their own experiences that allow them to recognize the right moments to make use of animal and vegetable resources. In this perspective, shadows constitute the places of attachment and affection, but they are also transformed into places of topophobia by the fact that they symbolize dark places and inhibiting phenomena, strange as when they serve as a shelter or touch of ferocious animals (wasps, snakes, among others), accumulation of feces, urine, or when they represent in general what is strange as the space of the deceased, of ghosts, spaces of curses, etc.

The solar diagramming allowed the deduction of cognitive narratives centred on the comparison of fragments of the corpus of interviews with fragments of the relevant literature. This methodological procedure was based on the interpretation of the social subjects' talks about how they mean everyday life from their life experiences with the meanings that science makes about academic projects.

From the reading and interpretation of the solar diagram it can be concluded that the solar altitude angle varies throughout the day, from sunrise around 5.30 am to sunset around 6.30 pm, constituting long days with about 14 hours of sunshine, making favourable plant and animal species that support this duration of days and nights. The variation in size and position of the shadows are different according to the shape of the objects that the projectiles and the seasons, with the solar map of the study areas coinciding with the position of shadows pointing east at the equinoxes, while in the winter solstices to the southeast and summer to the northwest. The sun has a slope ranging from ninety (90°) degrees to twenty-three (23)°degrees. Allowing us to deduce that the experiences of perception of the imaginary line and its phenomena from shadows, has a close relationship with scientific technological knowledge and can serve as a basis for the formulation of theories and concepts in academia.

REFERENCES

AFONSO, G. Myths and stations in the Tupi-Guarani sky. *Scientific American Brazil*, n. 45, p. 38- 47, 2006.

AGOSTINHO (Holy Bishop of Hippo). *Confessions*. São Paulo: Cultural April, 1987.

AKBARI, H.; ROSENFELD, A.; and TAHA, H. Sum@@mer heat islands, urban trees, and white surfaces. Atlanta-Georgia: American Society for Heating, Refrigeration, and Air Conditioning Engineers, 1990.

ALBUQUERQUE, U. P. *et al*. Methods and techniques for ethnobiological data collection. In: ALBURQUERQUE, U. P. *et al* (Org.). *Methods and techniques in ethnobiological and ethno-ecological research*. Recife: Nupeea, 2010.

ALMEIDA, Armando Antunes de. *Monografia agrícola* de *Massinga (Posto Sede)*: Memories of the overseas research council. Lisbon: [*s. n.*], 1959.

ALMEIDA, Voltaire de Oliveira. *Conceptual maps as instruments potentially facilitating the significant learning of concepts from the physical perspective*. 2006. 232f. Dissertation (Masters in Physics) - Universidade Federal do Rio Grande do Sul, Porto Alegre, 2006. Available at: http://www.lume.ufrgs.br/handle/10183/11794. Access on: 08 Mar. 2016.

ALVES, S. The mathematics of GPS. *Journal of the Mathematics Teacher*: RPM, n. 59, 2006. Available at: http://www.sinaldetransito.com.br/artigos/GPS_para_engenharia_%20de_%20transito.pdf. Access on: 19 Feb. 2016.

ALVES, Sérgio; CARVALHO, João Pitombeira; MILIES, Francisco César Polcino. *The geometry of the globe.* [*S.l.*]: IME/USP, 2009. Available at: http://www.bienasbm.ufba.br/M29.pdf. Access on: 20 Feb. 2016.

ARAUGH, Diones Charles Costa de. A proposal for the insertion of topics of Brazilian indigenous astronomy in high school: challenges and possibilities. 2014. 185f. Dissertation. (Master in Science Teaching) - Universidade de Brasília DF, Brasília, 2014.

AXT, Rolando; SILVEIRA, Fernando Lang da. An intriguing optical phenomenon: attraction between shadows. *Physics at school*, v. 8, n.1, 2007.

BACHELARD, Gastón. *The formation of the scientific spirit*: contribution to a knowledge analysis. Rio de Janeiro: Counterpoint, 2005.

BARION, Mariana Regina Lingiardi; SILVA, HeloísaCelis da; FERREIRA, Solange Gomes Colhado. The importance and types of shadows used for grazing cattle. *In*: INTERNAL SCIENTIFIC INITIATION WORKSHOPS, 6., 23 to 26 Oct. 2012, Maringá, PR. *Annals* [...].Maringá, PR: UniCesumar, 2012. Available at: http://www.cesumar.br/prppge/pesquisa/mostras/vi_mostra/mariana_regina_lingiardi_barion.pdf. Access on: 22 Dec. 2017.

BAXANDALL, Michael. *Shadows and lights*. São Paulo:EdUSP, 1997.

BELCHIOR, Elisabeth Morais. The *importance of shade in the green spaces of a transmontane city:* a case study. 2014. 48f. Dissertation (Masters in Forest Resource Management) - Instituto

Politécnico de Bragança, Bragança, 2014.

BENCHIMOL, Samuel, *et al.* Tropic and environment. In: TROPICOGeo-Bio-Social: proceedings of the Tropicology Seminar. Recife: Fundação Joaquim Nambuco; Massangana, 2002.

BIBLE of Jerusalem. São Paulo: Paulines, 1980.

BOBROWSKI, R. *Structure and dynamics of street forestation of Curitiba, Paraná, in the period 1984 - 2010.* 2011.144f. Dissertation (Masters in Forest Engineering) - Federal University of Paraná, Curitiba, 2011.

BORD,Richard J.; FISHER, Ann; O'CONNOR, Robert E. Public perceptions of global warming: United States and international perspectives. *ClimateResearch*, v. 11, p. 75-84, dez. 1998.

BRAZIL. *Supplementary Law No. 124* of January 3, 2007. Superintendence of Amazonian Development. Brasília: Presidency of the Republic, 2007. Available at: http://www.planalto.gov.br/ccivil_03/LEIS/LCP/Lcp124.htm. Access on: 23 Feb. 2016.

BRAZIL. Ministry of Education, Secretariat of Basic Education. *PCN+ High School*: educational guidelines complementary to the National Curricular Parameters: Nature Sciences, Mathematics and their technologies. Brasília: MEC; SEB, 2002.

BRAZIL. Ministry of the Interior. Banco da Amazônia S.A. *Preliminary economic diagnosis of the urban areas of Acre, Amapá, Roraima, and Rondônia.* Belém: Ministry of the Interior, 1969.

CÁRDENAS, G.; HARVEY, C. A.; IBRAHIM, M.; FINEGAN, B. Diversity and richness of birds in different habitats in a fragmented landscape inCañas, Costa Rica. *Agroforestry in the Americas,* v. 10, p. 78-85, 2003.

CARVALHO JUNIOR, Waldir deet *al.* Tropical soils: a vision according to free access world bases. *In*: BRAZILIAN SYMPOSIUM OF REMOTE SENSORY - SBSR, 17., 25-29 Apr. 2015,João Pessoa, PB. *Annals* [...]. João Pessoa, PB: INPE, 2015. Available at: https://www.alice.cnptia.embrapa.br/bitstream/doc/1019666/1/2015015.pdf. Access on: 28 Dec. 2017.

CARVALHO, M. M; XAVIER, D. F; ALVIM, M. J.; AROEIRA, L. J. *Sistemas Silvipastoris*: consortium of trees and pastures. Viçosa, MG: [*s. n.*], 2002.

CASATI, Roberto. *The discovery of the shadow*: from Platão to Galileo, the history of an enigma that fascinates humanity. São Paulo: Company of Letters, 2001.

CASTRO, H. S.; DIAS, T. C. Environmental Perception and Urban Afforestation in Macapá, Amapá. *Biota Amazônia*,Macapá, AP, v. 3, n. 3, p. 34-44, 2013.

CHANG, J. H. *Climate and agriculture*: an ecological survey. Chicago: Aldine, 1974.

CHARLET, Christian. *Le Père-Lachaise*: in the heart of the Paris of the living and the dead. Paris: Gallimard, 2003.

CHERMAN, Alexandre; VIEIRA, Fernando. *The time that time has*: why the year has 12 months and other curiosities and other curiosities about the calendar. 2. ed. Rio de Janeiro: ZAHAR, 2011.

CHOPRA, Deepak; FORD, Debbie; WILLIAMSON, Marianne. *The shadow effect.* [*S. l.*]: Moon

of
Paper, 2010.

CLAVAL, Paul. *Cultural geography*. Florianópolis: UFSC Publishing House, 2007.

CLAVAL, Paul. *Epistemology of geography*. Florianópolis: UFSC Publishing House, 2011.

CLAVAL, Paul. *Land of men*: geography. São Paulo: CONTEXTO Publishing House, 2014.

CONCEPT, Maristela Neves da. *Evaluation of the influence of artificial shading on the development of dairy heifers in pastures*. 2008.138f. Thesis (PhD in Agronomy) - Luiz de Queiroz College of Agriculture, University of São Paulo, Piracicaba, 2008.

CONTI, Bueno José. Geography and tropicality. *Revista da Casa da Geografia de Sobral*, v. 12, n.1, p.47-58, 2010.

CORBELLA, Oscar; YANNAS, Simos. *In search of a sustainable architecture for the tropics*. Rio de Janeiro: Revan, 2003.

COSTA, M. J. R. P. da; CROMBERG, V. U. Some aspects to be considered to improve the welfare of animals in rotated pasture systems. *In*: FOUNDATIONS of rotated grazing. Rev. Ed. of the Proceedings of the 14th Symposium on Pasture Management. Piracicaba, SP: FEALQ, 2005.

D'AMARAL, Marcio Tavares *et al. The time of the times*. Rio de Janeiro: Zahar, 2003.

DAYS SON, Moacyr Bernardino. Silvipastoral systems in the recovery of degraded tropical pastures. In: SYMPOSIES OF THE ANNUAL MEETING OF THE BRAZILIAN SOCIETY OF ZOOTECNIA, 43., 2006, João Pessoa. *Annals* [...]. João Pessoa: SBZ; UFPB, 2006. Special Supplement of the Revista Brasileira de Zootecnia, v.35, 2006.

FAVARETTO, Celso F. *Tropicália*: allegory, joy. Cotia, SP: Editorial workshop, 1996.

FETERIS, S.; HUTTON, D. Astronomy laboratory: what are we going to make today? *Publ. Astron. Soc. Aust.* , v.17, n.2, p. 116-118, 2000. Disponível em: http://www.publish.csiro.au/?act=view_file&file_id=AS00116.pdf. Acesso em: 9 mar. 2013.

FIGUEROA, Silvio N.;NOBRE, C. A. Precipitation distribution over central and western tropical South America. *Climanalise*, v. 5, n. 6, p. 36-45, 1990.

FRASER, A. F.; BROOM, D. M. *Farm animal behavior and welfare*. London: BalierreTindall, 1990.

FROTA, A. B.; SCHIFFER, S. R. *Thermal comfort manual*. São Paulo: Nobel, 1995.

FROTA, A. B. *Geometry of Sunstroke*.São Paulo: Geros, 2004.

GIL, Antonio Carlos. *Methods and techniques of social research*. 4. ed. São Paulo: Atlas. 1987.

GILSON, Etienne. *Introduction to the study of Saint Augustine*. Saint Paul: Editorial Discourse, 2006.

GODOY, Arllda Schmidt. Introduction to qualitative research and its possibilities. *Revista de Administração de Empresas*, São Paulo, v. 35, n. 2, p. 57-63, mar./abr. 1995.

GOLDENBERG, Mirian. *The art of research*: how to do qualitative research in Social Sciences.

8. ed. Rio de Janeiro: Record, 2004.

GOYAL, Megh Raj; BUILES, Victor Hugo Ramirez. *Elements of agroclimatology*. Risaralda: El Jazmin University Campus, [2015?]. Available at: https://www.researchgate.net/profile/Victor_Ramirez_Builes/publication/265739288_ELEMENT OS_DE_AGROCLIMATOLOGIA/links/5615cee308ae4ce3cc656448.pdf. Access on: 27 Dec. 2017.

HARVEY, David. *Postmodern condition*. 10. ed. São Paulo: Loyola, 2001.

HAYASHIADE, Juliana Midori, *et al*. Skin diseases among rural workers exposed to solar radiation: an integrated study between the areas of occupational medicine and dermatology. *Revista Brasileira de Medicina do Trabalho*, São Paulo, SP, v. 8, n. 2, p. 97-104, 2010.

HEAD, H. H. *Management of dairy cattle in tropical and subtropical*. [*S.l: s. n.*], 1995.

HENRIQUES, Giuliana Cristina César. *"Everything is medicine"*: study of healing practices in Maruanum, AP. 2011. Dissertation (Masters in Tropical Biodiversity) - Universidade Federal do Amapá, Macapá, 2011.

HERTZ, Hellen *et al*. Development of Moiré de Sombra Technique as a low cost alternative for postural analysis. *Scientia Medica*, v. 15, n. 4, p. 235-242, 2005.

BRAZILIAN INSTITUTE OF GEOGRAPHY AND STATISTICS (IBGE). *Population estimates for Brazilian municipalities on 01.07.2014*.[Brasília, DF], 2014. Available at: http://www.ibge.gov.br/home/estatistica/populacao/estimativa2014/estimativa_dou.shtm. Access on: 16 Feb. 2016.

NATIONAL STATISTICAL OFFICE(INE). Statistics of Massinga District, 2010-2012.Mozambique, 2012.Available at: http://www.ine.gov.mz/estatisticas/estatisticas-territorias-distritais/inhambane/marco-de-2012/distrito-de-massinga.pdf/view. Access on: 26 Feb. 2016.

INGOLD, Tim. From the transmission of representations to attention education. *Education*, Porto Alegre, v.33, n. 1, pp. 6-25, jan./abr. 2010.

INGOLD, Tim. Humanity and animality. In: INGOLD, T. (ed.). *Companion encyclopedia of anthropology*. London: Routledge, 1994a.

INGOLD, Tim. *The Perception of the Environment*. Essays on livelihood, dwelling and skill. London and New York: Routledge, 2000.

INGOLD, Tim. *What is an animal?* London: Routledge, 1994b.

NATIONAL INSTITUTE OF METEOROLOGY (INMET). *Manager's report*: financial year 2000. São Paulo: Ministry of Agriculture and Supply, 2001. Available at: http://www.inmet.gov.br/html/informacoes/relatorio_gestor/pdf/7DISME_REL-GESTOR_2000.pdf. Access on: 24 Feb. 2016.

JACKSON, E. Daytime astronomy in the northern hemisphere using shadows. *Astronomy Education Review*, v.2, n.2, sep. /jan. 2003/2004. Disponívelem: http://scitation.aip.org/getpdf/servlet/GetPDFServlet?filetype=pdf&id=AERSCZ00000 2000002000146000001&idtype=cvips. Acesso em: 26 dez. 2017.

JUNG, C. G. *Aion*:studies on self symbolism. 8. ed. Petrópolis: Voices, 2011.

LATOUR, B. *Reaggregating the social*: an introduction to the theory Actor - Network. Bauru, SP: EDUSC/ Salvador, BA: EDUFBA, 2012.

LUDKE, Menga; ANDRÉ, Marli E. D. A. *Research in education*: qualitative approaches. Rio de Janeiro: Pedagogical and University, 1986.

LIGHT, Buna. *Sunshine geometry*: apparent movement of the sun and use of the solar chart. [*S.l.*], 2015. [Discipline] AUT 272 - Sun, Architecture and Urbanism, Class 2, Solar Charts, 2. sem. 2015.Available   at: http://www9.fau.usp.br/arquivos/disciplinas/au/aut0272/Aula2_cartas_Solares.pdf. Access on: 26 Dec. 2017.

LYRA, R. Predominance of the wind in the coastal board region near Maceió, AL. *In*: CONGRESSO BRASILEIRO DE METEOROLOGIA, 10., 1998, Brasília. *Annals* [...]. Brasília, DF: [*n. º*], 1998.

MACHADO. Daniel Iria. Apparent movement of the sun, shadows of objects and measurement of time in the eyes of seventh grade students. *Latin American Journal of Astronomy Education - RELEA*,   n.   15,   p.   79-94,   2013.   Available   at: http://www.relea.ufscar.br/index.php/relea/article/viewFile/8/4. Access on: 20 Dec. 2017

MARENGO, C.; NOBRE, J. A. *Climate of the Amazon region*: weather and climate in Brazil. São Paulo: Workshop of Texts, 2009.

MARQUES, José Geraldo W. *Fishing fishermen*. São Paulo: NUPAUB, 2001.

MARTINS, S. R.; GONZALEZ, J. F. Evapotranspiration and physiological responses of beans grown in a greenhouse substrate with ventilation/heating system. *Revista Brasileira de Agrometeorologia*, v. 3, p. 31-37, 1995.

MATURANA, H. R. *et al*. The ontology of reality. Belo Horizonte: UFMG Ed., 2014.

MATURANA, Humberto. *Cognition, science and daily life*. Organization and translation by Cristina Magro and Victor Paredes. Belo Horizonte: UFMG, 2001.

MATURANA, Humberto; VARELA, Francisco. *The tree of knowledge*: the biological bases of human knowledge. Campinas: Ed. Psy, 1995. São Paulo: Palas Athena, 2012.

MEANS, Jefferson Duate Barros de. *Looking at the shadow through the ritualistic use of the sacred drink ayahuasca*. Monograph (Specialization in Psychology) - FACIS, Junguiana, Brasília, 2016.

MENDONÇA, F.; DANNI-OLIVEIRA, I. M. de. *Climatology*: Basics and Climates of Brazil. São Paulo: Oficina de Textos, 2007.

MILONE, André de Castro. *Astronomy in daily life*. Introduction to Astronomy and Astrophysics: activity booklet. São José dos Campos: INPE, 2002.

MILONE, André de Castro. *Astronomy in everyday life*. São José dos Campos: INPE, 2003.

MINAYO, Maria Cecília de Souza. *The challenge of scientific knowledge*: qualitative research in health. 2. ed. São Paulo/Rio de Janeiro: Hucitec-Abrasco, 1993.

MIOT, Luciane DonidaBartoli; MIOT, Hélio Amante; SILVA, Márcia Guimarães da; MARQUES, Mariângela Esther Alencar. Pathophysiology of melism. *Anais Brasileiros de Dermatologia*, v.84, n.6, nov./dec. 2009.

MONTEIRO, Karine Cristine Rodrigues; OLIVEIRA, Rosana Pena dos Santos de. Reflections on the consequences of verticalization for the urban climate in the city of Vitória da Conquista, BA, Brazil. In: LATIN AMERICA GEOGRAPHIC MEETING, 14., 2013, Peru. *Annals* [...]. Peru: [*n.*], 2013.

MORGAN, Gareth; BERGAMINI, Cecilia Whitaker; CODA, Roberto. *Images of the organization*. São Paulo: Atlas, 1996.

MORIN, Edgar. *Introduction to complex thinking*. Porto Alegre: Sulina, 2007.

MORIN, Edgar. *The seven knowledges necessary for the education of the future*. São Paulo: Cortez; Brasília, DF: UNESCO, 2001.

MUCHANGOS, Aniceto dos. *Mozambique*: landscapes and natural regions. Maputo: Author's Edition, 1999.

MUMFORD, Lewis. *The city in history*: its origins, transformations and perspectives. São Paulo: Martins Fontes, 1991.

NERI, S. H. A. *The use of geoprocessing tools to identify communities exposed to hepatitis A in the hangover areas of the municipalities of Macapá and Santana/AP*. 2004. 173f. Dissertation (Master's Degree in Civil Engineering/Water Resources) -Coordination of Post-Graduate Programs in Engineering, Federal University of Rio de Janeiro, Rio de Janeiro, 2004.

NEVES, Daniel Gonçalves *et al*. Regional climate modelling two years of precipitation extremes over Amapá: sensitivity test to convective schemes. *Revista Brasileira de Meteorologia*, v. 26, n. 4, p. 569-578, 2014.

NEWTON, Isaac. *Mathematical principles, optics and the weight and balance of* fluids. São Paulo: Editora Abril, Cultural, 1979.

NODA, S. N. *et al*. Landscapes and ethnoconomics in Ticuna and Cocama agriculture on the upper Solimões River, Amazonas. *Bol. Mus. Para. Emílio Goeldi. Cienc. Hum.* Bethlehem, v. 7, n. 2, p. 397-416, May/Aug. 2012.

NOWAK, D. J.; NOBLE, M. H.; SISINNI, S. M., DWYER, J. F. People and trees: assessing the US urban forest resource. *J. Forest*, v.99, p. 37-42, 2001.

NUNES, Felipe Santos de Miranda. Methodological proposal for assessing impacts, vulnerabilities and adaptation to climate change in Minas Gerais. [*S.l.*], 2012. Technical Note n. 2/2012 GEMUC/DPED/FEAM.

OLIVEIRA, A. S.; SANCHES, L.; DE MUSIS, C. R.; NOGUEIRA, M. C. J. Benefits of afforestation in urban squares: the case of Cuiabá, MT. *Rev. Elet. in Management, Education and Environmental Technology*, Cuibá, v. 9, n. 9, p. 1900-1915, Feb. 2013.

OMBE, Zacarias. *Dictionary of the main concepts*. Maputo, 2007.

OSMAN, Samira Adel; RIBEIRO, Olívia Cristina Ferreira. Art, history, tourism and leisure in the cemeteries of the city of São Paulo. *Licere*, Belo Horizonte, v. 10, n.1, p. 1-15, Apr. 2007. Available at: http://www.anima.eefd.ufrj.br/licere/pdf/licereV10N01_a6.pdf. Access on: 8 Nov. 2016.

PAIVA, Cauduro Dias de; CLARKE, Eloiza M.; Robin, T. Time trends in rainfall records in Amazonia. *Bulletinofthe American MeteorologicalSociety*, v. 76, n.11, p. 2203-2209, 2011.

DISTRICT OF MOZAMBIQUE ISLAND. *Strategic District Development Plan, 2008 - 2012*. Ilha de Moçambique, Nampula, Mozambique, 30 Oct. 2012.Available at: https://issuu.com/artpublications/docs/pedd_ilha_de_mo__com_del_incluido. Access on: July 19, 2017.

PEREIRA, Henrique dos Santos. The dynamics of the socio-environmental landscape of the Solimões-Amazonas River floodplains. In: FRAXE, Therezinha J.P.; WITKOSKI, Antônio Carlos (org.). *Amazonian riverside communities*: ways of life and use of natural resources. Manaus: EDUA, 2007.

PEREIRA, J. C.; CUNHA, D. de N. F. V.; CECON, P. R.; FARIA, E. S. Performance, rectal temperature and respiratory frequency of dairy heifers of three genetic groups receiving diets with different fibre levels. *Rev. Brasileira de Zootecnia*, v. 37, n. 2, p. 328-334, 2008.

PEREIRA, J.C.C. *Fundamentals of bioclimatology applied to animal production*. Belo Horizonte: FEPMVZ, 2005.

PEREIRA, R. A. *Expansion and urban planning in Macapá*: the case of Gleba Infraero. Monograph (Architecture and Urbanism Course) - Federal University of Amapá, Santana-AP, 2013.

PEREIRA, Silvia Regina Mendes. *Socio-health* repercussions *of the femur fracture epidemic on the survival and functional capacity of the elderly*. 2003. 164 f. Thesis (PhD in Public Health) - Oswaldo Cruz Foundation, Rio de Janeiro, 2003.

PLATON. *Timaeus*. Belém: EDUFPA, 2001.

POPIM, Regina Céliaet *al*. Skin cancer: use of preventive measures and demographic profile of a risk group in the city of Botucatu. *Science & Collective Health, p.* 1331-1336, 2008.

PORTILHO, Ivone dos Santos. *Urban development policies in segregated spaces*: an analysis of PDSA in the city of Macapá (AP). 2006. 166f. Dissertation (Masters in Geography) - Universidade Federal do Pará, Belém, 2006. Available at: http://ppgeo.propesp.ufpa.br/ARQUIVOS/dissertacoes/2004/DISSERTAÇÃO%20IVONE%20P ORTILHO.pdf. Access on: 22 Dec. 2017.

PRIMAVESI, Ana Cândidaet *al*. Nutrients in the phytomass of marandu grass depending on sources and doses of nitrogen. *Science and Agrotechnology*, v. 30, n. 3, p. 562-568, 2006a.

PRIMAVESI, Ana. *Ecological soil management*: agriculture in tropical regions. São Paulo: Nobel Prize, 2006b.

RAGON, Michel. *L'espace de lamort*: essaisurl'architecture, la décoration et l'urbanismefunéraires. Paris: A. Michel, 1981.

RAUPP, Carlos F. M. *The climate of the earth*: processes, changes and impacts. São Paulo: University of São Paulo, 2010.

RIBEIRO, C. A. M. *Application of geoprocessing techniques for analysis of the relationships between the factor of vision of the sky and the different orientations of the urban network*. Monograph (Technology in Geoprocessing) - Instituto Federal de Educação, Ciência e Tecnológica da Paraíba, João Pessoa, 2003.

RICE, R.A.; GREENBERG, R. Silvopastoral systems: ecological and socioeconomic benefits and migratory bird conservation. *In*: SCHROTH, G.; FONSECA, G.A.B; HARVEY, C.A.; GASCON, C.; VASCONCELOS, H.L.; IZAC, A-M.N. (Ed.) *Agroforestry and Biodiversity Conservation in Tropical Landscapes*. Washington: Island Press, 2004. p. 453-472.

RIPADO, Mário Fernades Bento. The Machongos of Inharrime and Inhambane regions: contribution to their study. *Mozambique*, n. 62, p. 5-60, trim. apr./ jun. 1950.

RISTÓTELES. *Physics*. Madrid: Editorial Gredos, 1995.

RODRIGUES JÚNIOR, Manoel Alves. *The calendars and his contribution to the teaching of astronomy*. 2012. Dissertation (Masters in Physics and Astronomy) - University of Porto Department of Physics and Astronomy, Porto, 2012.

RODRIGUES, A. L., SOUZA, B. B. D.; PERREIRA FILHO, J. M. Influence of shading and cooling systems on thermal comfort of dairy cows. *Agropecuária Científica no Semiárido*, v. 6, n. 2, p. 14-22, 2010.

ROHDE, G. M. *Environmental Epistemology*: a philosophical-scientific approach to allopoietic human effect. Porto Alegre: EDIPUCRS, 1996.

SANTOS, Adilson. A tour of the double territory. *Investigations magazine*, v. 22, n. 1, 2009.

SANTOS, Milton. *The nature of space*. São Paulo: Editora HUCITEC, 1996.

SCANDIUZZI, P. P. Indigenous *education x indigenous school education*: an ethnocidal relationship in ethnomathematical research. Marilia: UNESP, 2000.

SERGIA, Rafael Augustus. The concept of social representation in the works of Denise Jodilet and Serge Moscovici. *Years 90*, Porto Alegre, n.13, jul. 2000.

SILVA, Fernando Siqueira da; CATELLI, Francisco; GIOVANNINI, Odilon. A model for the apparent annual movement of the sun from a geocentric perspective. *Cad. Bras. Ens. Fís.*, v. 27, n. 1, pp. 7-25, Apr. 2010.

SILVA, J.G.R. Orbital cycles or Milankovitch cycles. *In*: WINGE, M. *et al*. Illustrated Glossary texts. [*S.l.*], 2007. Available at: http://sigep.cprm.gov.br/glossario/index.html. Access on: 28 Dec. 2017.

SILVA, Mário Adelmo Varejão. Meteorology and climatology. *Digital version*, Recife, v. 2, mar. 2006.

SILVA, Roberto G. Prediction of the shading configuration of trees in cattle pastures. *Agricultural Engineer*, Jaboticabal, v. 26, n. 1, p.268-281, jan./abr. 2006. Disponível em: https://repositorio.unesp.br/bitstream/handle/11449/27779/S0100-69162006000100029.pdf?sequence=1&isAllowed=y. Access on: 28 Dec. 2017.

SNOW, Frank J. American Society Of Heating, Refrigeration, And Air Conditioning Engineers (ASH RAE) Thermographic Standard 101 P. In: Thermal Infrared Sensing Applied to Energy Conservation in Building Envelopes, 1982. p. 94-98.

TAVARES, Ana Paula Cunha; TOSTES, José Alberto. The urban evolution of a city in the middle of the world. *Revista Nacional de Gerenciamento de Cidades*, v. 1, n. 4, p. 36-41, 2013.

TAVARES, John Paul Nardin. Characteristics of the climatology of MacapáAP1. *Caminhos de geografia Umberlândia*, v.15, n.50, p. 138-151, jun. 2014.

TODOROV, J.C.; HENRIQUES, M.B. What is not and what could be behaviour. *Revista Brasileira de Análise do Comportamento*, v. 9, n. 1, p. 74-78, 2013.

TODOROV, Tzvetan. *Theories of Symbolism*. Brazil: Unesp Editions, 2013.

TOMÁS DE AQUINO. *Compendium of Theology*. São Paulo: Editora Abril, Cultural, 1979.

TONNEAU, F. Behavior and the skin. *Revista Brasileira de Análise do Comportamento*, v. 9, n. 1, p. 66-73, 2013.

TORRINHA, Mário Nunes. Urban food supply networks and spatial organisation: the supermarket network in MacapáAP. Supervisor: Heleniza Ávila Campos.
2013. 170 f. Thesis (PhD in Regional Development) - University of Santa Cruz do Sul (UNISC), Santa Cruz do Sul, RS, 2013.

TROGELLO, Anderson Giovani. *Learning objects*: a didactic sequence for teaching astronomy, 2013. 102 f. Master (in Science and Technology Teaching) - Universidade Tecnológica Federal do Paraná (UTFPR), Ponta Grossa, PR, 2013. Available at: http://repositorio.utfpr.edu.br/jspui/bitstream/1/1251/1/PG_PPGECT_M_Trogello%2C%20Ander son%20Giovani_2013.pdf. Access on: 8 Jan. 2016.

TUAN, Yu-Fu. *Topophilia*. São Paulo: Difel, 1980.

TURNBULL, C.M. *The forest people*. London: Chatto e Windus. People of África. New York: Holt, Rinehart & Winston, 1965, p.128-310.

VALLADARES, Clarival do Prado. *Art and society in Brazilian cemeteries*. Brasília, DF: National Press, 1972.

REERJAN-SILVA, M.A. *Meteorology and Climatology*. Brasília: MAP, 2001.

VARELLA, Irineu Gomes. The sun is apine in your city. *CoelumAustrale*: Personal newspaper of astronomy, physics and mathematics, year 4, n. 29, feb. 2013.

VENTURE, Magda Maria. The case study as a research modality. *Rev. Socerj*. v. 20, n.5, p. 383-386, 2007. Access on: 23 Nov. 2015.

VOVELLE, Michel. *The hour of the great passage*: chronicle of death. Paris: Gallimard, 1993.

YIN, Robert K. *Case study*: planning and methods. Porto Alegre: Bookman, 2010.

ZWEIG, C; ABRAMS, J (org). *Meeting the shadow*: the cult potential of human nature. São Paulo: Cultrix Publishing, 2011.

Buy your books fast and straightforward online - at one of world's fastest growing online book stores! Environmentally sound due to Print-on-Demand technologies.

Buy your books online at
www.morebooks.shop

Kaufen Sie Ihre Bücher schnell und unkompliziert online – auf einer der am schnellsten wachsenden Buchhandelsplattformen weltweit! Dank Print-On-Demand umwelt- und ressourcenschonend produziert.

Bücher schneller online kaufen
www.morebooks.shop

KS OmniScriptum Publishing
Brivibas gatve 197
LV-1039 Riga, Latvia
Telefax: +371 686 204 55

info@omniscriptum.com
www.omniscriptum.com

Printed by Books on Demand GmbH, Norderstedt / Germany